Smart Polymers and Composites

Edited by Abu Nasar

Polymeric compounds are generally blended with inorganic/organic materials to prepare composites to tailor the desired properties for specific requirements. The present book reviews new research in the fields of composite green polymers for environmental applications, polyaniline based composites for wastewater treatment, smart polymeric coating materials, polymer decorated bimetallic nanosorbents for dye removal, fuel cell materials, polymeric membranes, green bio-nanocomposites and polymer based catalysts.

Smart Polymers and Composites

Edited by
Abu Nasar

Department of Applied Chemistry, Faculty of Engineering & Technology,
Aligarh Muslim University, Aligarh, India

Published by **Materials Research Forum LLC**
Millersville, PA 17551, USA

Published as part of the book series
Materials Research Foundations
Volume 21 (2018)
ISSN 2471-8890 (Print)
ISSN 2471-8904 (Online)

Print ISBN 978-1-945291-46-3
ePDF ISBN 978-1-945291-47-0

Distributed worldwide by

Materials Research Forum LLC
105 Springdale Lane
Millersville, PA 17551
USA
http://www.mrforum.com

Manufactured in the United States of America
10 9 8 7 6 5 4 3 2 1

Table of Contents

Table of Contents

Preface

Materials science is an interdisciplinary field of science and engineering covering the properties of materials and their applications in various fields. Polymeric compounds are generally blended with inorganic/organic materials to prepare composites to tailor the desired properties for specific requirements. The present edition is an outcome of significant contributions of specialists from different interdisciplinary areas.

This book includes the topics related to composite green polymers for environmental applications, polyaniline based composites for wastewater treatment, smart polymeric coating materials, polymer decorated bimetallic nanosorbents for dye removal, fuel cell materials, polymeric membranes, green bio-nanocomposites and polymer based catalyst for dehydrogenation of dimethyl ammonia borane.

I am grateful to all the contributing authors/coauthors for their contributions and the publisher as well. The editor is also thankful to the Chairman, Department of Applied Chemistry, Faculty of Engineering & Technology, Aligarh Muslim University, Aligarh for providing the necessary facilities.

Chapter 1

Production of Composite Green Polymers and Their Effects on the Environment

Santanu Sarkar[1], Chiranjib Bhattacharjee[2]*, Supriya Sarkar[1]

[1] Environment Research Group, R&D, Tata Steel Ltd., Jamshedpur-831007, India

[2] Chemical Engineering Department, Jadavpur University, Kolkata-700032, India

* cbhattacharyya@chemical.jdvu.ac.in, c.bhatta@gmail.com

Abstract

Nowadays several disadvantages regarding the application of synthetic composites encourages the use of green composite polymers instead. The main disadvantages like disposal problem, non-biodegradability are fully resolved by using natural fiber reinforced polymers. Green composites consist of both biodegradable natural reinforcement and a polymer matrix. Biocomposites are mainly used in the automotive and construction sectors. Research regarding the development of green composites is ongoing globally in the polymer science to improve efficiency and applicability of these materials. Moreover, the use of green composites started the reuse of agricultural waste. In this chapter, the sources of different types of green fibers and natural oil based polymers have been discussed. Further, several methods for the preparation and application of thermoplastic and thermosetting bio-based composite polymers are reviewed.

Keywords

Composite Polymers, Natural Fibers, Agricultural Wastes, Natural Oil, Biodegradable

Contents

1.   Introduction

The word composite refers to a material that is made of two or more constituent materials with considerably different physical or chemical properties. Composite polymers have two major parts one is a strong load carrying material, known as reinforcement and the other weaker material is known as a matrix where the reinforcement material is embedded inside it. Basically, the matrix binds all reinforcements maintaining a certain structure so that the composite becomes rigid and it can withstand a large load. Moreover, the physical and chemical properties of each component of composite never change during the formation of the composite material and it is quite obvious that the individual components do not have similar strength as the composite [1]. In recent years, the uses of composite materials are mainly in the area of packaging, building materials, and commodities, as well as in hygiene products and almost all are manufactured as by-products of the petroleum industries. A study has estimated that the petroleum reservoirs will be exhausted in the next 50 years and moreover the USA, a single nation, produces 60 billion pounds of plastic waste, mostly originated from the petroleum industries [2]. These type of composite are non-biodegradable, therefore those are mainly used in lowland filling. They create obstacles in the drainage system, are non-recyclable, use high energy consumption during its production and are hazardous to health when inhaled. To reduce the detrimental effect of synthetic polymer to the environment biodegradable polymers have been proven as a better alternative. Biodegradable polymers are generally produced from plants, animals and microbes by means of biochemical reactions. Therefore, these types of polymers are easily degradable and termed as "*green composite polymers*". Starch, wheat gluten, whey protein, soy protein etc. are frequently used to manufacture polymers of this category [3-7].

A green composite combines plant fibers with natural resins to create composite materials. In 1980, the first green composite, melamine-formaldehyde resins reinforced with cotton or paper was introduced to the polymer market [8]. Natural fibers are emerging as a low-cost, lightweight and apparently environmentally superior alternative to synthetic fibers. After disposing of biodegradable polymers they are degraded in the presence of enzymes from living microorganisms. In the above process, initially, polymers are fragmented into lower molecular weight molecules by means of different actions such as oxidation, photodegradation, hydrolysis and microbial reaction etc. and finally microbes mineralize by-products from polymers. The whole degradation process depends on the sources and chemical structure of composite polymers as well as the environment for biodegradation [9]. Moreover, the different physical and chemical properties of polymers depend on the compositions and several techniques which are involved in polymer processing [10-14].

Nowadays, superior use of green composites is observed compared to synthetic polymers due to its recyclability, sustainability and eco-friendly nature [15, 16]. A good number of researchers are involved in the development of new technologies for the production of such green composites. Moreover, a major interest is to produce this type of composites from agricultural wastes, thus, making them economically compatible as well as safe for the environment [15, 16].

2.   Polymers versus composite green polymers

Any material is said to be green only when it is easily renewable and biodegradable and according to that, green polymers or composites are environmentally degradable without any external resources and easily recyclable. Moreover, after degradation, it does not produce secondary pollutants. Therefore, the uses of green polymers are advantageous considering every aspect of the environment. The main challenge is to find green and biodegradable polymer matrices for the production of composite polymers. Starch, lignin, cellulose acetate, poly-lactic acid (PLA), polyhydroxyalkanoates (PHA), polyhydroxylbutyrate (PHB), etc. are the common sources of natural polymers whereas, aliphatic and aromatic polyesters, polyvinyl alcohol, modified polyolefin etc. are synthetic polymers. Afforested said polymers are degradable and termed as biopolymers [15, 17]. However, the mentioned synthetic polymers are not renewable and biodegradable. Therefore, natural polymers must be used as matrix for green composites.

Glass fibers reinforced (GRP) is frequently used in different aspects of our daily life such as automotive, construction, electro/electronics, and sports [18]. To increase the mechanical strength and use in aviation, wind power, and sports the reinforcement of glass is replaced by carbon or polymeric fibers. Due to the high density of glass material, the weight of the GRP is also higher than other composite materials and therefore it cannot be used for light weight application. Moreover, during disposal glass fibers up to 50% by volume remains in residues thus, it cannot be termed as a green polymer though the matrix is completely degradable. Therefore, a better alternative for reinforcement in a green polymer matrix is needed. In the last 20 years, several researchers and industries have used plant fibers, mainly residues of vegetables, to produces fully green composite polymers. These types of polymers are light in weight and completely biodegradable or renewable. Furthermore, as these are produced from waste materials the production cost is cheap. Due to complete biological degradability of natural fibers reinforced composites (NFCs) are termed as *biocomposites* and nowadays, these types of composites are being used not only commercially but also industrially.

In a literature survey by La Mantia et al. [19] was described that NFCs can be used frequently if it is possible to remove process complicacy and improve the performance of it. Most of the polymers from by-products of petroleum are used in our daily life and day by day, the price of products from petroleum is rising. Both limiting sources and disposal problem are triggering the need for green polymers as an alternative. The term '*green composites*' refers to such composites which are produced from renewable resources and after disposal, degrade without leaving any impact on the environment [19]. This chapter aims to describe an updated general overview on green composites, their applications and production.

3.   Components of green composites

As already mentioned, both matrix and reinforcement of the green composite should belong to natural resources. The available sources of matrix and fibers are discussed in the following subsections.

3.1   Natural reinforcements: their properties and applications

The sources of natural fibers are divided into two main categories one is plant based and the other is animal based natural fibers. Detail classifications are shown in Fig. 1 and the physical properties of those fibers are summarized in Table 1.

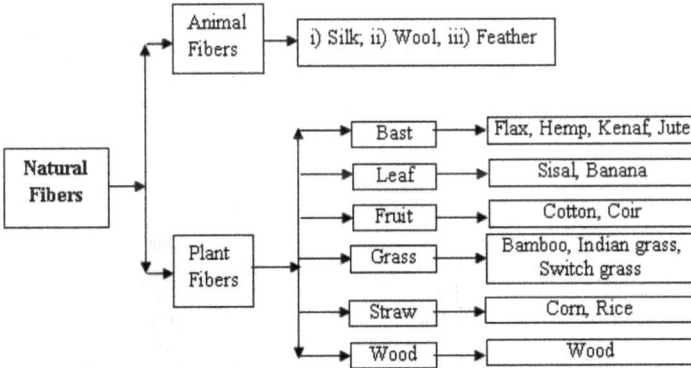

Fig. 1 Classification of natural fibers.

The composition of plant and animal cells are very different therefore, the vegetable fibers belong to polysaccharides whereas animal fibers are constituted with proteins. Both fibers are not readily available but are present in the bodies of plants and animals along with hemicelluloses, lignin, wax, proteins, etc. One needs to separate those ingredients from the fiber for reinforcement and there are some pretreatment processes for the purification of fibers. The details of physical characteristics of different fibers which are frequently used to prepare composites are represented in Table 1. Moreover, in the same table pretreatment process for different fibers and clear comparison of different physical properties between natural and synthetic fibers are indicated. Though the tensile strength of synthetic reinforcements is better than natural fibers, the density indicates that the natural ones are much lighter than synthetic fibers. Therefore, to produce lighter weight composites, natural fibers are the best alternative and such type of composites can be used for automotive application. Flax is a natural fiber with higher tensile strength near to glass reinforcement. Moreover, all natural fibers having good insulating properties, are produced from natural resources with the help of some minor pretreatment processes which are biodegradable, renewable and non-toxic in nature. Thus, a wide range of green composite polymers with different physical characteristics can be manufactured using such types of fillers.

Table 1 Some important physical properties of different synthetic and natural reinforcements with pre-treatment process.

Fiber source	Treatment	% Cellulose Content	Tensile Strength MPa	Elastic Modulus GPa	% Elongation	Specific tensile strength	Density (g/cm³)	Reference
E-glass	-	-	2000–3500	70	2.5	800–1400	2.5	[20]
Aramide	-	-	3000–3150	63–67	3.3–3.7	2140–2250	1.4	[20,21]
Carbon	-	-	4000	230–240	1.4–1.8	2860	1.4	[20]
Banana	NR	NR	779	32	2	NR	NR	[22]
Cotton	Alkaline treatment	NR	NR	82	13.57	190-530	1.5-1.6	[20,21, 23,24]
Bark of cotton stalks	NaOH treatment	79	377	18.7	3.0	NR	NR	[25]
Sisal	NR	NR	600–700	38	2.3	390-490	1.3-1.5	[20,21, 24,26]
Flax	NR	NR	800–1500	60-80	1.2–1.6	230-1000	1.5	[20,24,26]
Hemp	NR	NR	550–900	70	1.6	370-600	1.5	[24,26]
Jute	NR	NR	400–800	10-30	1.8	300-610	1.3-1.5	[20,21, 24,26]
Coir (coconut husk)	NR	NR	220	6	1.5–2.5	110-180	1.2	[20,21, 26]
Velvet Leaf	Alkaline treatment	69	325–500	18-38	1.6-2.6	NR	NR	[27]

Switch grass leaves	Simple alkaline treatment	61.5	715	31	2.2	NR	NR	[27]
Switch grass stems	Simple alkaline treatment	68.2	351	9.1	6.8	NR	NR	[29]
Stinging nettle	NR	NR	1594	87	2.11	NR	NR	[29]
Kenaf	Alkaline treatment	NR	130	11	1.3	NR	NR	[30]
Ramie	Untreated	68.6–76.2	560	24.5	2.5	270-620	1.5	[20,24,31]
Soybean straw	Alkaline treatment	85	351	12	3.9	NR	NR	[32]
Flax	NR	NR	1339	54	3.27	NR	NR	[33]
Wheat straw	Mechanical processing	47.55	58.7	3.7	NR	NR	NR	[34]
Wheat straw	Microbial retting	NR	139.9	4.8	NR	NR	NR	[35]
Wheat straw	Chemical processing	63.14	146	7.9	NR	NR	NR	[36]
Hop stem	Alkaline treatment	84	53.3	20	3.3	NR	NR	[36]
Cornstalks	Alkaline treatment	52	286	16.5	2.2	NR	NR	[37]
Soft wood kraft	NR	NR	1000	40	NR	670	1.5	[20]
Chicken feathers	Alkaline treatment	NR	100–200	3–10	NR	112–220	0.89	[8, 38]
Silkworm silk	Alkaline treatment	NR	NR	0.5	15	NR	1.3–1.4	[39-42]

* NR – Not reported

3.1.1 Plant reinforcement

As per the classification, those fibers belonging to plants are termed as plant reinforcements. According to Table 1, cellulose is the most effective fiber that can be used to prepare green composites. Moreover, it can be produced with different agricultural wastes and therefore the use of such type of fibers makes the process as well as composite polymers more economically viable. Due to good mechanical properties, low density and safer handling compare to synthetic fibers, it can be frequently used as a green reinforcement. Many researchers have found green fibers, especially vegetable fibers such as flax, hemp, jute, and kenaf, as effective due to their availability, strong mechanical property, ease of separation from the agricultural waste (Table 1). This table also indicates a wide range of variation of mechanical properties of vegetable fibers and that variation is greatly affected by the age of the plant, geographical location and climatic change, harvesting techniques, the process of purification or separation, etc. Accordingly the availability of bast fiber varies with a geographical location for cultivation consequently, e.g. flax and hemp are available in temperate regions whereas huge production of jute and kenaf is observed in tropical regions [43]. As already mentioned cellulose is a key component of plant fibers but there are two major problems existing in using such type of fibers; one is its quality changes according to geographical location and another is its hydrophilic property due to the presence of hydroxyl group in cellulose. Thus special care needs to be taken during the preparation of composite polymers so that it cannot absorb water. Furthermore, hydrophilic reinforcement is not compatible with the hydrophobic matrix, sometimes fibers are segregated from the matrix and therefore, poor mechanical strength is achieved by the composite. Above all, huge availability, low cost, good mechanical strength and simple processing techniques make successful use of plant reinforcements for green polymer composites.

As shown in Fig. 1, different types of bast fibers are easily available and the use of those fibers as reinforcements increases the value of these particular crops. Some plants are self-seeding or can be planted and can grow on non-fertile land or only require a small amount of fertilizer. Zou et al. [44] have found that Switchgrass stems could be used as reinforcement to produce composite polymers, which could be used industrially in the field of automotive interiors. Moreover, it was also found that not only having same density in the composite, reinforcement with switchgrass stem had higher modulus, higher flexural strength, higher impact resistance than jute composite but sound absorption properties of both composites are quite similar. In a similar type of study Liu et al. [45] has investigated that raw Indian grass as reinforcement in a composite polymer made of soy-based aliphatic–aromatic copolyester has shown improved tensile and flexural properties but did not impact strength. However, the tensile and impact properties of the soy-based resin were significantly enhanced in the presence of alkali treated Indian grass.

In the early 90's the use of hemp fiber as reinforcement in polypropylene-based composites has increased significantly [46]. This type of fiber could be purified with the help of steam explosion in a flash hydrolysis laboratory pilot unit. Then the surface treatment was done using propylene maleic anhydride copolymer to enhance the performance of hemp fibers, which can be used as reinforcement in polypropylene. As hemp is a plant fiber, therefore it is consist of 55% cellulose. The strength of the composite was increased with the increase of filler concentration and the composites with treated fibers showed better performance because of good adhesion property between fillers and matrix. The percentage of elongation and tensile modulus contradicted each other as the first one decreased with increase in fiber content in the polymer and a reverse situation was observed for the latter case. A similar research work [47] has shown that the presence of 20% of bamboo fibers in the propylene matrix increases 12.5 and 10 % of flexural and tensile strengths respectively.

Wambua et al. [26] used different types of plant fibers i.e. sisal, kenaf, hemp, jute and coir fibers as reinforcement in polypropylene based green composites and the reinforcement was done with help of compression molding using a film stacking method. The physical characteristics of the mentioned composite like elastic modulus, failure strain and tensile strength were measured from the stress-strain curve. Instron tensile test machine helped to perform flexural tests using the three-point bending technique and impact tests were carried out using an Instron Charpy impact testing machine. The results from different analyses have been provided in Table 1. It has been observed that hemp fiber composites have highest tensile strength of 52 MPa whereas coir has the lowest about 10 MPa and mechanical strength of other three composites made of kenaf, sisal and jute were the same i.e. approximately 30 MPa. The most significant thing is that hemp composites have a flexural strength of 54 MPa which is nearly the same as glass mat composites of 60 MPa. It was observed that the tensile modulus of hemp and kenaf composites was greater than coir composites. From the above comparisons of mechanical properties of natural fiber composites to the glass material composites, it may be concluded that natural fibers can be used as an alternative reinforcement.

A similar study was carried out to prepare the polyester resin, as the matrix of the composite using chemically modified banana fiber as reinforcement [22]. The chemical treatment was done using two types of silanes (A174 and A151) and 40% fibers were used to make green composites with polyester. Mechanical properties of the fiber as well as composites were examined with the help of dynamic mechanical analyzer and scanning electron microscope (SEM) confirmed micro-structural properties of all composites. Thermal treatment was required to improve the mechanical properties of the fibers.

The potential application of wheat straw fibers has been investigated by Panthapulakkal et al. [35]. After their chemical and mechanical treatment, wheat straw was used as reinforcement in the preparation of polypropylene composite. The fibers were treated with NaOH and refined mechanically. Moreover, the composite with mechanically treated fibers showed 29% and 49% improved tensile and flexural strength whereas those mechanical properties with chemically treated fibers showed only 12% and 49% improvement. Better performance was observed for the chemically processed fiber than the mechanically treated fibers, as chemically treated fibers were more homogeneous, stronger and longer than the other type of fibers. Moreover, poor dispersion of the cellulose present in chemically treated fibers resulted in a strong interaction between the hydroxyl groups of cellulose.

In another study, with variation of 0–40 wt.% ramie cellulose nanocrystallites (RN) was used as fillers to produce environmental friendly glycerol plasticized starch (PS) composites [48]. Acid hydrolysis process was adopted to prepare RN, having an average fiber length of 538.5 ± 125.3 nm and diameter of 85.4 ± 25.3 nm and the morphological characteristics of generated fibers were carried out using SEM, differential scanning calorimeter (DSC), thermogravimetric analysis (TGA), dynamic mechanical thermal analysis and tensile testing. The above-mentioned protocols showed that synergistic interactions between fillers as well as between filler and PS matrix were responsible for reinforcement during the production of the composites. Tensile strength and Young's modulus of the composite material containing 0–40 wt% RN could be improved with the help of controlled humidity treatment of 50% relative humidity and therefore, those mechanical properties increased from 2.8 MPa to 6.9 MPa and from 56 MPa to 480 MPa respectively. However, unfortunately, it was observed that the elongation at break decreased from 94.2% to 13.6%.

In 2007, Bodros et al. [49] prepared various types of composite materials using different types of matrices such as Polylactic acid (PLA), L-polylactide acid (PLLA), poly 3-hydroxylbutyrate (PHB), polycaprolactone and starch thermoplastic, poly butylene succianate (PBS) and poly butylene adipateco terephtalate (PBAT) and flax fiber as reinforcement with the help of the film stacking technique. Under the same tensile loading condition, tensile properties of composites were compared with polypropylene-flax composites. Young's modulus of PLLA polymer was measured 3321 MPa and that value was improved to 9519 MPa in the presence of 30% flax fiber by volume fraction in PLLA-flax composite. Moreover, 55% improvement of ultimate stress was also observed and the strain to failure reduced from 2.4% to 1.5%.

Singha and Thakur [50] reported synthesis and characterization of different types of bio-composite polymers. In which they used *Hibiscus sabdariffa* fiber as a reinforcing material in urea–formaldehyde (UF) resin based polymer as a matrix. During the production of composite reinforcement was done using particle, short and long fibers of *Hibiscus sabdariffa* and their optimizations were performed. Tensile, compressive and wear properties were measured as a function of fiber loading with the help of a computerized Universal Testing Machine and morphological analysis of composites was performed using SEM. The best mechanical properties were observed when particle reinforcement was done compared to other reinforcement. Moreover, during the comparison loading capacity, it was observed that 332.8 N at the extension of 2.2

mm, 307.6 N at the extension of 2.23 mm and 286.1 N at the extension of 2.28 mm respectively for those composite reinforced with the particle, the short and long fiber of *Hibiscus sabdariffa*. Not only loading capacity but similar observations were found during the testing of compressive strengths and wear resistances. Those observations were explained with the help of SEM images and it was revealed that there was more uniform mixing particle reinforcement and matrix inside the composite as compared to short and long fibers. This type of environment friend composites can be used for a variety of industrial applications.

Cellulose nanofibers (CNF) were used as reinforcement for the production of mango puree-based edible films for packaging applications [51] with a variation of CNF concentration up to 36 % (wt/wt). Three types of properties of composite materials i.e. tensile properties, water vapor permeability, and glass transition temperature were tested so far according to standard protocols. Incorporation of CNF in composite polymers increased tensile strength and Young's modulus up to 8.09 MPa and 322.05 MPa respectively. According to the authors, at higher concentration of CNF cross-linking took place between the fibers and the matrix. Therefore, water vapor permeability decreased with 10% (wt/wt) CNF loading and the glass transition temperature was improved significantly.

Kenaf fiber was used to prepare a polylactic acid resin-based unidirectional fiber reinforced composite [52]. The composite polymer was fabricated with kenaf fibers bundles having 50–150 μm diameters at the temperature of 180 °C. According to the standard mechanical properties testing the tensile strength of the fabricated composite decreased when kept under loading condition for 60 minutes. A linear relationship between fiber content up to 50% and tensile or flexural strengths or elastic modulus was observed. At 70% fiber content composite had the maximum value of tensile strength of 223 MPa and flexural strengths of 254 MPa. Recently, normal cellulose from different sources can be used for the production of the composite. Among those cellulose fibers having the length of 2.1 mm and a width of 9.7 μm from the bark of cotton stalks were used in a research work [25] and in that work, the fibers are separated from cotton stalks with the help of an alkaline treatment. The composites were prepared using those cellulose fibers in polypropylene matrix using compression molding. After preparation tensile properties were tested on an Instron tensile tester and it was found that the composites had flexural strength of 12.4 MPa, offset yield of 22.5 MPa, stiffness of 1.4 N/mm, modulus of elasticity of 502 MPa, tensile strength of 15.7 MPa, tensile modulus of 806 MPa and an impact resistance of 0.3 J. A further investigation on natural cellulose fibers from milkweed stem was performed by Reddy [53]. The alkaline treatment was used to obtain the cellulose fibers from milkweed stems. The amount of cellulose contained in fibers was measured as per the protocol of AOAC method and as per the result, 75.4% cellulose was present in 0.9 mm average length single cells and diameter of 13.0 µm. The percentage crystallinity was tested using an XRD and the fiber had a crystallinity of 39%. The morphological analysis was done with help of SEM images. Tensile properties were measured by using the Instron tensile testing machine and according to that fibers had a breaking elongation of 4.7% and Young's modulus of 15.8 GPa. As the physical properties of the milkweed stem fibers are quite similar to cotton and linen, it can be used in textile, composite, automotive and other fibrous applications.

Famá et al. [54] identified the potential use of wheat bran as fillers in the production of biodegradable composites where cassava starch containing glycerol and potassium sorbate was used as matrix. The filler concentration was varied in three different ways such that 1.5 mg, 13.5 mg and 27.1 mg/g of the matrix. With the increase of filler concentration, the mechanical properties of the composite were improved as well as the storage modulus and hardening of the films increased because of higher water-insoluble fiber content in wheat bran. Moreover, due to the same reason, composites showed a decrease in the moisture content and in presence of wheat bran the composite had an improved water vapor barrier property. However, with an increase of filler contain beyond 70% the tendency of deformation and rupture was enhanced.

Kaushik et al. [55] recently established a pathway to use waste agricultural cellulose fibers to produce a nanocomposite using natural polymers as a matrix. The novel green nanocomposite was prepared using plasticized corn starch, which was a thermoplastic starch (TPS), and nanofibers prepared from wheat straw. The composite was prepared using following steps. Firstly, the nanofibers were extracted using steam explosion and acidic treatment and it followed high shear mechanical treatment. After that, the generated nanofibers were dispersed by varying its concentration in TPS using a high shear mixer. Thereafter it was necessary to measure different properties of the nanocomposites and according to these measurements; the maximum tensile modulus of 220 MPa was achieved with yield strength of 6.5 MPa at 15% fiber content. The SEM image confirmed the presence of nanofibers and a typical SEM image of nanocomposites is shown in Fig. 2.

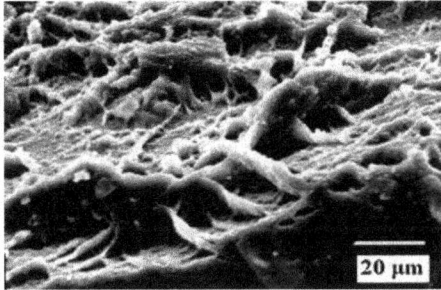

Fig.2 A typical SEM image of starch cellulose nanocomposites.

Qu, *et al.* [56] prepared nanocomposites with solvent casting methods from N, N-Dimethylacetamide (DMAC) and using Polylactic acid (PLA) as a matrix and cellulose nanofibers obtained from bleached wood pulp. Both chemical and mechanical treatments were used to prepare cellulose nano-fibrils and thereafter, those were dispersed in organic solvent. Polyethylene glycol (PEG) was added to the PLA to increase the interfacial bonding between the matrix and the nanofibers. The suurprising thing was that in the presence of nanofibers in the PLA matrix the tensile strength and percentage of elongation were 30 MPa with 2.5% elongation respectively, which was lower than same values in the case of pure PLA. Thus, to improve the mechanical properties PEG was added PLA and cellulose nanofibrils. As a result, the tensile strength and the elongation of nanocomposite increased by 28.2% and 25%, respectively, compared with pure PLA polymer, as well as those values increased by 56.7% and 60% compared with the PLA-cellulose nanocomposite. From the FTIR analysis, it was observed that there was an improvement in intermolecular hydrogen bonding interaction in the presence of PEG.

In the year of 2010, Liu et al. [57] made a successful attempt to produce starch composites reinforced by bamboo cellulosic crystals (BCCs) which were prepared using an acid treatment. First bamboo fibers were treated with HNO_3–$KClO_3$ and after that, those were hydrolyzed using H_2SO_4. Glycerol plasticized starch was used as matrix. SEM, XRD and NMR were used to study morphological characteristics and according to that, BCCs consisted of cellulose particle with 50-100 nm in diameter. The mechanical properties were measured using Instron Tensile machine and as per the testing result, the elongation at break increased with the decrease of BCC content. It was found that the composite of plasticized starch plastic with 8% of BCCs loading had a higher efficiency than any other compositions.

Banana and sugarcane fibers are generally thrown out after taking the usable part from them. Guimarães, et al. [58] tried successfully to use those waste materials as reinforcement for the production of composite materials where they used starch containing 28% amylose as a matrix. That type of matrix was prepared from corn while fibers and reinforcements were obtained using chemical as well as mechanical treatments from the banana pseudostem and sugarcane bagasse. Thereafter, the mixing of fillers and polymers were done with the help of a ball mill. XRD and Instron tensile testing machine were used to check the crystallinity of both starch & its composites and to obtain stress-strain curves respectively. The present study indicated that the matrix consisted of 70 wt.% starch and 30 wt.% glycerol which, had good thermal stability. The banana fibers contain varied in three different ways i.e. 20%, 25% and 35% and at those particular value Young's modulus increased by 186%, 294% and 201% respectively whereas, the tensile strength remained relatively constant. Moreover, at the same percentage of banana fibers, the yield strength (YS) increased by 129%, 141% and 133% respectively.

Sometimes different types of fibers which are treated as waste material like sunflower stalk [59], bagasse [59], rice husk [60], wheat straw [61, 62], soy stalk [62], and cornhusk [63] work as valuable sources of cellulosic fibers and those have been successfully used as reinforcement in green composite polymers.

3.1.2 Animal fiber reinforcement and their applications

In earlier section, the potential applications plant fibers for reinforcement in green composite have been elaborately discussed. Now a day's not only plant fibers but animal fibers are used for reinforcement though the numbers of animal fibers are less compare to plant fibers. In general, animal fibers are used by the textile industries to manufacture garments and commonly used animal fibers are wool [64] and silk [40, 65]. Moreover, those have also been used as reinforcements for the production of green composites. Wool is a keratin fiber and it has higher surface toughness, flexibility, high aspect ratio, and is less hydrophilic than plant fibers, which belong to cellulose fibers. It has been investigated that a feather has a hollow structure and is mainly made of keratin fibers containing a certain volume of air [66]. Therefore, it has a very low density of 0.9 g/cm^3 and the low-dielectric constant of 1.7 and thus composites made of feathers can have potential application for electronic applications [66]. The unit of silk from cocoons of the silkworm is a single continuous silk strand having the length of 1000–1500 m and the component is fibroin filaments, which are cemented on silk by sericin. Different properties of silk are variable and those properties are greatly affected by the silkworm species and the speed spinning during the formation of silk [41]. Higher spinning speed favors producing stronger and more brittle silk, whereas the reverse case produces weaker and more extensible fibers. Some properties of silk i.e. oxidation resistance, antibacterial and UV-resistant are noticeable but it has lower thermal stability. Silk fibers reinforced composite has a wide range of application in bioengineering.

3.2   Green polymers – matrix in green composites

As discussed earlier a green composite polymer is made of both a biodegradable matrix and reinforcement. Green polymers are derived environmentally from biological resources for the preparation of green composites. According to ASTM D6866 the carbon content can be calculated, as a percentage of weight of the total carbon content with respect to the amount of isotope ^{14}C present in that particular material. In general, a ^{14}C isotope of carbon is not present in the material, which is fully based of fossil carbon. Whereas materials, those belong to partially or entirely based on renewable resources have a certain percentage of ^{14}C isotope content. Now a day's, several bio-based natural thermosetting and thermoplastic polymers are available and those can be used to produce green composites. The major constitutions of green polymers and their availability in the current market have been provided in Table 2. The descriptions of several types of green polymers and their applications, as well as the methods of application, are discussed in the following subsections.

Table 2 Major constituents of green polymers and their availability.

Polymers	*Green* constituent	*% of Green* constituent	Availability
Polyhydroxyalkanoates (PHAs)	Polymer itself	100	Commercial
Polylactic acid (PLA)	Lactic acid	100	Commercial
Polytrimethylene terephthalate (PTT)	1,3-Propanediol	30	Commercial
Polyacrylic acid	3-Hydroxypropionic acid	100	Non-commercial
Polyethylene (PE)	Ethanol	100	Commercial

Polybutylene succinate (PBS)	Succinic acid	55	Non-commercial
Polyvinylchloride (PVC)	Ethanol	100	Commercial
Polyamide 11 (PA11)	Ricinoleic acid	100	Commercial
Cellulosics	Cellulose	100	Commercial
Thermoplastic starch (TPS)	Starch	100	Commercial
Polyisoprene	Isoprene	100	Commercial
Aliphatic polyesters	1,3-Propanediol, succinic acid, fatty acids	100	Commercial
Polyethers	1,3-Propanediol, fatty acids	100	Commercial
Polyurethane (PU)	Fatty acids	70	Commercial
Furan resins	Xylose	100	Commercial
Epoxy resins	Triglycerides	100	Commercial
Polyamide (6/10)	Ricinoleic acid	65	Commercial

3.2.1 Thermoplastic starch based composites

De Carvalho et al. [67] first used starch to produce a thermoplastic composite with the help of melt intercalation in a twin-screw extruder. The composites were made of corn starch plasticized with glycerin as matrix and hydrated kaolin as reinforcement. To study the effect of plasticizers in structural as well as mechanical properties, Pandey and Singh [68] varied the sequence of addition of plasticizers to the matrix. They observed the significant effect of the sequence of addition of components during the formation of composites but no such critical observation was found during the production of the composite. Thereafter, to produce a new green composite with cellulose fibers as reinforcement, Guan and Hanna [69] modified the starch with acetate. They varied different parameters such that degrees of substitution (DS) of 1.68 and 2.3 of starch

16

acetates prepared using 70% amylose cornstarch along with 10%, 20% and 30% (w/w) cellulose and 20% (w/w) ethanol. Moreover, a twin screw extruder was used with the variation of barrel temperatures of 150, 160 and 170 °C as well as screw speed of 170, 200 and 230 rpm. However, Kumar and Singh [70] did a new modification of starch by means of photo-induced cross-linking and that modification was carried out by following the aqueous dispersions of starch. They used the microcrystalline cellulose as reinforcement and glycerol as a plasticizer to produce the composite. At the same time sodium benzoate as photosensitizer was introduced in the composite via casting and after that, it was irradiated with ultraviolet (UV) light. Moreover, several research works using starch with a variation of experimental procedure and parameters have been carried out by Lu *et al.* [48], Ma *et al.* [71], Svagan [72], Famá *et al.* [54], Kaushik *et al.* [55], Liu *et al.* [57], Guimarães *et al.* [58] and Kaith Avella et *al.* [74]. These studies indicate that starch can be used to produce green composites.

3.2.2 Poly lactic acid based composites

Ogata *et al.* [75] used polylactic acid (PLA) for the fabrication of composite polymers. PLA along with organically modified clay was dissolved in hot chloroform in the presence of modified dimethyl distearyl ammonium with montmorillonite (MMT) (2C18MMT). There are several nanocomposites using PLA prepared by Sinha *et al.* [76-81], Yamada *et al.* [82], Maiti *et al.* [83], Paul *et al.* [84], Lee *et al.* [85] and Chang *et al.* [86]. These studies have shown that bio-composites prepared with PLA as a matrix have better mechanical and thermal properties. Bondeson and Oksman [87] used cellulose whiskers to reinforce a commercial PLA matrix for prepareing PLA composite. Lee *et al.* [88] prepared PLA composites with improved mechanical properties by melt compounding and injection molding. Qu, *et al.* [56] modified commercial grade PLA with a poly ethylene glycol (PEG) as a compatibilizer to use as a matrix and reinforced it with cellulose fibrils, produced from wood pulp by bleaching. In such case, PEG enhanced the interfacial bonding between PLA and cellulose fiber. In another way, the PLA composites were fabricated by Lee *et al.* [88] using solvent casting methods in presence of N, N-Dimethylacetamide (DMAC). Therefore, PLA has potential application in the preparation of nanocomposites.

3.2.3 Cellulose based composites

Agricultural products are the best source of cellulose and can be used a great source for polymer. The 20 wt.% of triethyl citrate (TEC) used as a plasticizer with 80% cellulose acetate (CA) to produce green composites in presence of organically modified clay by melt compounding [89]. The cellulose-based composites have very important application in recent trends like humidity and temperature sensor [90]. Moreover, this type of composite was developed using methylcellulose (MC) as a matrix with MMT as reinforcement [91], which has active antimicrobial property and could be used as packaging material. Further research works by Zimmermann *et al.* [92] and Zadagan *et al.* [93] revealed that cellulose based nanocomposite with hydroxylapatite could be used medical applications. Therefore, above studies showed that cellulose based composites have potential applications in different sensitive and valuable areas.

3.2.4 Plant oil based composites

The epoxidized soybean oil, diglycidyl ether of bisphenol-A (DGEBA) and organically modified MMT were used to prepare green bio-based composite polymers [94]. Modified MMT clay with variation of its concentration from 1 wt.% to 7 wt.% was mixed with soy-based epoxy resin by a mechanical stirrer to produce composite material and hence the composite had improved tensile strengths. Azeredo *et al.* [51] developed a composite using mango puree-based edible films for packaging application and for the production of this type of composite cellulose nanofibers (CNF) in different concentrations 36 wt.% was used as nano-reinforcement. In another process for the production of bio-based nanocomposites soy-based polyurethane was used as a matrix [96]. Moreover, polyurethane matrix can be modified with Halloysite Nanotubes (HNT) to produce E-glass reinforced composites and consequently, the shear strength increased from 14.28MPa for 0 wt.% HNT to 21.28MPa for 2.4 wt.% HNT. Low-cost vacuum assisted resin transfer molding method was used for fabrication of such type of composites.

3.2.5 Polymer–polymer blends based composites

It is obvious that if two or more bio-degradable polymers are mixed together the resultant polymer should be biodegradable and hence this type of resultant polymers can be used for the production of green composites with improved degradability and mechanical properties. Different research groups have used several types of polymer blends like Starch/PLA blends, polybutylenes succinate/cellulose acetate blends, starch/modified polyester blends, polycarprolactone/polyvinyl alcohol blends and thermoplastic starch/polyesteramide blends [97-102] to produce composites. PLA, polycaprolactone (PCL) and TPS were used by Sarazin *et al.* [103] to produce binary or tertiary blends using a one-step extrusion process. Sometimes, natural rubber was mixed with starch to form bio-based composites [104].

3.2.6 Other biopolymers based composites

This category of biopolymers polyhydroxyl butyrate (PHB), gelatin and chitosan can be used to produce bio-composites. PHB is a polyester, which is synthesized by several bacteria in the environment. Gelatin can be produced by thermal denaturation of collagen isolated from animal skin and bones with dilute acid and chitosan can be extracted from exoskeletons of crustaceans, insects and from the cell wall of fungi as well as microorganisms [105-108].

It has been observed that in the current scenario, there is high demand for bio-based plastics in the market and globally the growth rate for the use of such type of polymers is gradually increasing. As petrochemical sources are limited, the prices of petroleum by-products are gradually increasing due to high demand in the market. Therefore, proper economic development, strategies, and technologies are required to utilize bio-based polymer resources. Only then the use of green polymers become economically advantageous and can be used as a replacement for petroleum-based polymers as well as in the application of several aspects in life [36].

At present, there are only a few green polymers beeing commercialized to use as matrices for the production of green composites which have been indicated in Table 2. In next subsequent sections, the production of green composites and their potential applications have been elaborately discussed.

4.   Green composites – production and applications

Green composites are derived from renewable resources with improved mechanical properties, which depend on fiber aspect ratio, volume fraction and orientation, and on the adhesive force between fiber and matrix [109]. Those values varied with the change of plant fibers [110]. In not only the plant fibers but this type of changes were observed in animal fibers also e.g. feather keratin and silkworm silk fibers have an average diameter of 5 μm [8] and 10–15 μm [40] respectively. The properties of fibers depend on extraction process and the suitable matrix needs to be identified so that the adhesion force should be the maximum between matrix and reinforcement. Several processes for the production of the green composites are mentioned in literature [19, 20, 111, 112-116]. First, fibers are extracted and modified with help of different process e.g. esterification [117-122], etherification [117, 121, 122], treatment with silane (SiH$_4$) [111, 112, 123], isocyanates [124, 125], plasma treatment [111, 113, 126, 127] corona treatments [126, 128]) and similarly some modifications of the polymer matrix can be carried out according to requirement [129, 130].   With help of chemical treatment and grafting method, the polarity of the fiber could be changed according to the polarity of the matrix to improve the attraction between matrix and fibers. Consequently, PEG grafted on flax fibers to produce PLA composites having improved mechanical properties [121]. A similar type of research work had been carried out for the copolymerization of vinyl monomers (methyl methacrylate and acrylates) of vegetable and animal fibers [131, 132].

The same purpose can be fulfilled with the addition of a compatibilizer which is widely used in NFCs with polyolefin matrices [19]. According to that, Wong et al. [133] used glycerol triacetate and 4, 4-thiodiphenol to improve interfacial adhesive force between flax fibers and PLA or PHB.

4.1　Thermosetting green composites

The resins from natural oils from soybean, castor, linseed, etc. could be used as an alternative of thermosetting resins from petrochemicals e.g. vinyl esters, and epoxy resins [134]. Moreover, availability and cost of production of natural oils make those lucrative for the production of green composites. The polymerization of triglycerides of fatty acids unsaturated plant oils takes place after fictionalization of active sites i.e. double bonds, allelic carbons, ester groups, carbons alpha to the ester group and such type of polymerization can be done with epoxidation, ring opening reaction with haloacids or alcohols, ozonolysis, and hydration etc. [134]. Nowadays green composite polymers are used in agricultural equipment, automotive components, infrastructure (bridges and highway components), marine structures, rail infrastructure (carriages, box cars, and grain hoppers), and construction industry (particleboard, ceilings, and engineered lumber) and the details of thermosetting green composite polymers are listed in Table 3. Lee et al. [124] used the mixture of polyol from castor oil and polycaprolactone diol as matrix and isocyanate modified hemp fibers to reinforce the polyurethane green composites. Acrylate epoxidized oils can be prepared from the reaction between epoxidized natural oils and acrylic acid. Similarly, to fabricate low dielectric polymer Hong and Wool [66] used avian fibers and acrylate epoxidized soybean oil resin and this type of composite has an application in high-speed microelectronics. The "*closed mold*" techniques e.g., resin transfer molding, vacuum infusion, and compression molding is very common practice to produce thermosetting green composites [83] with high fiber loading. Initially, the resin is infused into fibers in a mold and thereafter, the curing of the composite matrix is done. The curing process is temperature sensitive and that can be performed by following compression molding or vacuum infusion. Sometimes compression molding and vacuum infusion may be applied alternatively. Though several techniques are available for the production of thermosetting polymers, it is lacking behind in wide range application due to insufficient supply for thermosetting polymer matrix in the market. However, ample amount of soybean oil based resin is available in the market. A German company Jakob Winter is one of the manufacturing industries of GreenLine, composite products made of flax and hemp fibers as reinforcement in epoxidized natural-oil resins by compression molding. According to the manufacturer, these biocomposites can be varnished and laminated. A USA based company Environ Biocomposites Mfg. produces green composites using soybean resin and waste fibers i.e. recycled newsprint, wheat straw, and sunflower hulls and those composites are used for indoor applications.

4.2 Thermoplastic green composites

Thermoplastic green composites are mainly prepared from commercial thermoplastic polymers PLA and the major source of PLA is starch. Graupner et al. [142] found out that NFCs is the most effective fibers that can be used as reinforcement to produce PLA green composites. The details of the major components of thermoplastic green composites are given in Table 3. According to Barkoula et al. [148], PLA polyhydroxyalkanoates (PHAs) has a major application in the production of bio-composites. Table 3 indicates that in the case of thermoplastic composites the maximum fiber content is 50 wt.%. The most significant thing for thermoplastic green composites is that this type of composite polymers can be produced with the help of compounding and injection molding methods and those can be done quite easily by means of some standard equipment commonly used in the plastic industry. Only short fibers can be used for the reinforcement, which is the main limitation of compounding and injection molding, though for longer fibers compression molding can be introduced. Thermoplastic composite polymers are used mainly in the automotive industry [110].

There are many companies like Jakob Winter (Germany), FKuR (Germany), Kareline Oy (Finland), GreenGran BV (Germany), FASAL WOOD KG (Austria), FuturaMat (France) that produces PLA based green thermoplastics using flax/hemp fibers, wood fibers, Kareline PLMS fibers, wood or natural fibers, wood waste, wood fibers as reinforcement. Most of the cases wood waste is used as reinforcement and composites are used in automotive interiors due to its thermal insulation property.

Table 3 Types of thermosetting and thermoplastic green composites and their constituents.

Fibers	% of Fibers	Matrix	Process	Types	Ref.
Chicken feather	5–20 wt%	Acrylate epoxidized soybean oil resin	Vacuum-assisted resin transfer molding	Thermosetting	66
Hemp	5 wt%	Mixture of polyol from castor oil and polycaprolactone diol	Resin transfer molding	Thermosetting	124
Luffa Cylindrica	10 wt%	Castor oil with diphenylmethane diisocyanate	Sheet molding compound	Thermosetting	135

Flax	60 wt%	Metacrylated soybean oil with styrene	Compression molding	Thermosetting	136
Jute	50 wt%	Polyurethane and epoxy resins from Mesua Ferrea seed oil	Compression molding	Thermosetting	137
Hemp	0-65 wt%	Epoxidized linseed oil with methyl tetrahydrophthalic anhydride	Compression molding	Thermosetting	132
Wheat straw	50-90 wt%	Linseed oil, maleic anhydride, and divinylbenzene	Compression molding	Thermosetting	61
Flax	0-15 wt%	Epoxidized soybean oil	Compression molding	Thermosetting	138
Coconut, sisal	15-30 wt%	Castor oil with diphenylmethane diisocyanate	Compression molding	Thermosetting	139
Kenaf	10 wt%	Epoxidized soybean oil with maleic anhydride	Resin transfer molding	Thermosetting	123
Flax	20-40 wt%	Acrylated epoxidized soybean oil with styrene	Resin transfer molding	Thermosetting	141
Flax	20-30 wt%	PHB, thermoplastic starch, PLA	Compression molding	Thermoplastic	49
Wood flour	0-30 wt%	Polyhydroxybutyrate (PHB)	Injection molding	Thermoplastic	143
Pineapple leaf	0-30 wt%	Poly(3-hydroxybutyrate-co-3-hydroxyvalerate) (PHBV)	Compression molding	Thermoplastic	144
Wood	10-40 wt%	PHBV	Injection molding	Thermoplastic	145
Wood	10-30 wt%	Thermoplastic starch	Injection molding	Thermoplastic	146
Wood flour	15-60 wt%	Thermoplastic starch	Melt mixing	Thermoplastic	147
Flax	25% (v/v)	PHBH	Injection molding	Thermoplastic	122
Flax	8-40% (v/v)	PHB, PHBV	Compression molding and injection molding	Thermoplastic	148

Straw	0-50 wt%	PHB	Injection molding	Thermoplastic	73
Wood	20-40 wt%	PHB, PHBV	Injection molding	Thermoplastic	39
Flax, ramie, jute, cellulose fiber, and oil palm fiber	10-25 wt%	PHBV, Thermoplastic starch blends, cellulose acetate	Injection molding	Thermoplastic	95
Corn straw, wheat straw, soy stalk	10-40 wt%	PHBV	Injection molding	Thermoplastic	62
Wood	0-30 wt%	Thermoplastic starch	Injection molding	Thermoplastic	146
Wool	5-20 wt%	Cellulose acetate	Solution casting	Thermoplastic	150
Hemp	0-30 wt%	Cellulose acetate	Compression molding and injection molding	Thermoplastic	149

5.   Area of applications and future prospects

In last twenty years, several types of research have been carried out in the field of green polymer composites but the commercial use of those composite materials are still limited. Though a huge amount of fillers are available the higher cost of polymer matrix and production cost of the composites are responsible for the present situation. Therefore, efficient process development and new sources for matrices are required for effective commercialization of these green composites.

In 2003, green composites of PLA reinforced with kenaf fibers have been commercially used by Toyota to cover the spare tire of a car. Due to light weight, better crash absorbance, better acoustic, and heat insulation PLA composites are used for automotive interior applications e.g. seatback lining, package shelves, seat bottoms, seatback cushions, head restraints, under-floor body panels, door panel inserts, armrests, etc. Moreover, thermoplastic green composites can be used in making toys, funeral articles, packaging materials, musical instruments, and electronic devices. Therefore, in 2006, modified PLA based composites developed by UNITIKA LTD were used by the NEC Corporation for the production of mobile phone parts. Presently, several companies use the green composites for mobile phone production.

Different parts of buildings i.e. terrace flooring, outdoor decking, window and door frames, panels for both indoor and outdoor uses can be made using wood reinforced thermosetting green composites [151]. Presently most the significant area of application of wood reinforced composites is in the furniture industries and the well-known company IKEA has already been commercializing chairs and shelves made of such composites [110, 151].

6. End disposal

It is important that after their use, the composite materials should be degraded without affecting the environment. By definition green composites are made of bio-based materials and more precisely the degradation of composites should be done biologically by following some international protocols. These protocols are the measurement scale of pollution level encompassed by the by-products of green composites after their degradation. Generally, the natural fibers used as reinforcements are biodegradable but the biodegradability of the composites depends entirely on the polymer matrix which is either biodegradable or not. The degradability of different matrices is schematically illustrated in Fig. 3. Green composites with a biodegradable polymer matrix do not have any disposal problem but the non-biodegradable matrix of green composite exhibits disposal problems hence, disposal of such materials is limited to incineration or landfilling. Biodegradability is not the only property of green composites but their better durability is also a requirement for building and furniture applications.

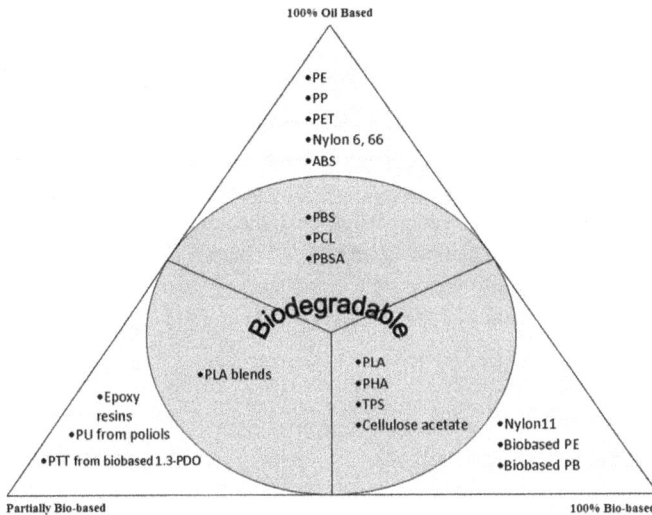

Fig. 3 Schematic representation of degradability of different polymer matrices.

The manufacturers of green composites are encouraged to reuse and recycle their products. Several actions such as EU-Directive 2000/53 (End of life vehicles - Waste - Environment - European Commission) for the automotive sector and the International Maritime Organization's Convention for the Safe and Environmentally Sound Recycling of Ships for the marine sector promote the detrimental effects of disposal of composite materials.

Nowadays researchers are trying to replace non-biodegradable components with biodegradable green composites to reduce the need for recycling. The joint research work between Ohio State University and the University of Akron (Goodyear Polymer Centre) has developed a 90% biodegradable car where all parts were made of bio-based polymers and fibers from renewable feedstock. At the same time, the first prototype of a ''green car'' was developed at the University of Warwick in 2009 and most surprisingly was that this car was a Formula 3 racing car. The project NavEcoMat developed by PoleMer Bretagne (France) together with companies and research centers aimed to produce high-performance green composites to build small boats. The flax fibers reinforced PLA was used to build a kayak prototype made for seawater [153].

7. Conclusions

There are several advantages present in use of green composites such as the light weight of composites enhances the sustainability and biodegradability removes the disposal problems. The production of such composites improves the usability of the bio-based waste materials. As raw materials of composites are easily available and therefore composite polymers are, in some cases, cheaper than synthetic polymers. Moreover, reuses as well as recycle of green composites make them more economically viable. The wide range variation of mechanical properties of composites can be obtained with a variation of the proportion of reinforcement and the polymeric matrix; therefore, green composites have a wide range of applications.

References

[1]   D. Hull, T.W. Clyne. An introduction to composite materials. Cambridge University Press, Cambridge, 1996. https://doi.org/10.1017/CBO9781139170130

[2]   E.S. Stevens, Green Plastics, Princeton University Press, Princeton, USA, 2002.

[3]   K. S. Miller, M.T. Chiang, J.M. Krochta. Heat curing of whey protein films, J. Food Sci., 62 (1997) 1189–1193. https://doi.org/10.1111/j.1365-2621.1997.tb12241.x

[4]   P. Lodha, A.N. Netravali. Characterization of interfacial and mechanical properties of green composites with soy protein isolate and ramie fiber, J. Mater. Sci., 37 (2002) 3657–3665. https://doi.org/10.1023/A:1016557124372

[5]   S. Nam and A. N. Netravali, Interfacial and mechanical properties of ramie fiber and soy protein "green" composites, ICCE-9, San Diego, California's (2002) 551-552.

[6]   A. Gennadios, C.L. Weller. Edible films and coatings from wheat and corn proteins, Food Technol., 44 (1990) 63–69.

[7]   L. Krull, G. Inglett. Industrial Uses of Gluten, Cereal Sci. Today, 16 (1971) 232–236.

[8]   J.R. Barone, W.F. Schmidt, C.F.E. Liebner. Compounding and molding of polyethylene composites reinforced with keratin feather fiber, Compos. Sci. Technol., 65 (2005) 683-692. https://doi.org/10.1016/j.compscitech.2004.09.030

[9]   N. Lucas, C. Bienaime, C. Belloy, M. Queneudec, F. Silvestre, J.E. Nava-Saucedo. Polymer biodegradation: mechanisms and estimation techniques, Chemosphere 73 (2008) 429-442. https://doi.org/10.1016/j.chemosphere.2008.06.064

[10] S. Kalia, K. Georgios, A. Silva, S. Furtado. Applications of green composite materials, Biodegradable Green Composites, John Wiley & Sons, Inc, (2016) 312-337. https://doi.org/10.1002/9781118911068

[11] K.L. Pickeringa, M.G.A. Efendya, T.M. Le, A review of recent developments in natural fibre composites and their mechanical performance, Composites Part A: Applied Science and Manufacturing, 83 (2016) 98-112. https://doi.org/10.1016/j.compositesa.2015.08.038

[12] L. Mohammed, M. N. M. Ansari, G. Pua, M. Jawaid, M. Saiful Islam, A review on natural fiber reinforced polymer composite and its applications, Inter. J. Polym. Sci. 2015 (2015) 1-15., Article ID 243947, doi:10.1155/2015/243947.

[13] O. Faruk, A. K. Bledzki, H-P. Fink, M. Sain. Progress report on natural fiber reinforced composites, you have free access to this content, Macromol. Mater. Eng., 299 (2014) 9–26. https://doi.org/10.1002/mame.201300008

[14] D. Briassoulis. Mechanical behaviour of biodegradable agricultural films under real field conditions, Polym. Deg. Stab., 91 (2006) 1256-1272. https://doi.org/10.1016/j.polymdegradstab.2005.09.016

[15] J.K. Pandey, W.S. Chu, C.S. Lee, S.H. Ahn. Preparation characterization and performance evaluation of nanocomposites from natural fiber reinforced biodegradable polymer matrix for automotive applications. Presented at the International Symposium on Polymers and the Environment: Emerging Technology and Science. Bio Environmental Polymer Society (BEPS), Vancouver, WA, USA, 17–20th October 2007.

[16] S.R. Sinha, M. Bousmina. Biodegradable polymer/layered silicate nanocomposites. In Polymer Nanocomposites; Woodhead Publishing and Maney Publishing: Cambridge, England, 57-129.

[17] M.J. John, S. Thomas, Biofibres and biocomposites, Carbohyd. Polym., 71 (2008) 343-364. https://doi.org/10.1016/j.carbpol.2007.05.040

[18] E. Witten. Composites Market: Market Developments, Challenges, and Chances, AVK Federation of Reinforced Plastics, Germany (2010).

[19] F.P. La Mantia, M. Morreale. Green composites: A brief review, Compos. A. Appl. Sci. Manuf., 42 (2011) 579-588. https://doi.org/10.1016/j.compositesa.2011.01.017

[20] A.K. Bledzki, S. Reihmane, J. Gassan. Properties and modification methods for vegetable fibers for natural fiber composites, J. Appl. Polym. Sci., 59 (1996) 1329-

1336. https://doi.org/10.1002/(SICI)1097-4628(19960222)59:8<1329::AID-APP17>3.0.CO;2-0

[21]   A.K. Mohanty, M. Misra, G. Hinrichsen. Biofibres, biodegradable polymers and biocomposites: An overview. Macromol, Mater. Eng., 1 (2000) 276-277. https://doi.org/10.1002/(SICI)1439-2054(20000301)276:1<1::AID-MAME1>3.0.CO;2-W

[22]   L.A. Pothan, S. Thomas. Polarity parameters and dynamic mechanical behavior of chemically modified banana fiber reinforced polyester composites, Compos. Sci. Technol., 63 (2003) 1231-1240. https://doi.org/10.1016/S0266-3538(03)00092-7

[23]   L.F. Zemljic, P. Stenius, J. Stana-kleinschek, V. Ribitsch. Characterization of cotton fibers modified by carboxymethyl cellulose, Lenzinger Berichte, 85 (2006) 68-76.

[24]   H.Y. Cheung, M.P. Ho, K.T. Lau, F. Cardona, D. Hui. Natural fibre-reinforced composites for bioengineering and environmental engineering applications, Compos. B: Eng., 40 (2009) 655-663. https://doi.org/10.1016/j.compositesb.2009.04.014

[25]   N. Reddy, Y. Yang. Properties and potential application of natural cellulose fibers from the bark of cotton stalks, Bioresource Technol, 100 (2009) 3563-3569. https://doi.org/10.1016/j.biortech.2009.02.047

[26]   P. Wambua, J. Ivens, I. Verpoest. Natural fibers: Can they replace glass in fiber reinforced plastics?, Compos. Sci. Technol., 63 (2003) 1259-1264. https://doi.org/10.1016/S0266-3538(03)00096-4

[27]   N. Reddy, Y. Yang. Characterizing natural cellulose fibers from velvet leaf (Abutilon theophrasti) stems, Bioresource Technol., 99 (2008) 2449-2454. https://doi.org/10.1016/j.biortech.2007.04.065

[28]   N. Reddy, Y. Yang. Natural Cellulose fibers from switchgrass with tensile properties similar to cotton and linen, Biotechnol. Bioeng., 97 (2007) 1021-1027. https://doi.org/10.1002/bit.21330

[29]   E. Bodros, C. Baley. Study of the tensile properties of stinging nettle fibers (Urtica dioica), Mater. Lett., 62 (2008) 2143-2145. https://doi.org/10.1016/j.matlet.2007.11.034

[30]   S.K. Batra. Other long vegetable fibers. In Handbook of Fiber Science and Technology. Marcel Dekker Fiber Chemistry: New York, NY, USA, 4 (1998) 727.

[31]   K. Goda, M.S. Sreekala, A. Gomes, T. Kaji, J. Ohgi. Improvement of plant based natural fibers for toughening green composites—Effect of load application during mercerization of ramie fibers, Compos. Part A Appl. Sci. Manuf., 37 (2006) 2213-2220. https://doi.org/10.1016/j.compositesa.2005.12.014

[32]   N. Reddy, Y. Yang. Natural cellulose fibers from soybean straw, Bioresource Biotechnol., 2009 (100) 3593-3598.

[33]   C. Baley. Analysis of the flax fiber tensile behavior and analysis of the tensile stiffness increase, Compos. Part A Appl. Sci. Manuf., 33 (2002) 939-948. https://doi.org/10.1016/S1359-835X(02)00040-4

[34]   M. Sain, S. Panthapulakkal. Bioprocess preparation of wheat straw fibers and their characterization, Ind. Crops Products, 23 (2006) 1-8. https://doi.org/10.1016/j.indcrop.2005.01.006

[35]   S. Panthapulakka, A. Zereshkian, M. Sain. Preparation and characterization of wheat straw for reinforcing application in injection molded thermoplastic composites, Bioresource Biotechnol., 97 (2006) 265-272. https://doi.org/10.1016/j.biortech.2005.02.043

[36]   N. Reddy, Y. Yang. Properties of natural cellulose fibers from hop stems, Carbohyd. Polym., 77 (2009) 898-902. https://doi.org/10.1016/j.carbpol.2009.03.013

[37]   N. Reddy, Y. Yang. Structure and properties of high quality natural cellulose fibers from corn stalks, Polymer, 46 (2005) 5494-5500. https://doi.org/10.1016/j.polymer.2005.04.073

[38]   D.N. Saheb, J.P. Jog. Natural fiber polymer composites: a review, Adv. Polym. Technol., 18 (1999) 351-363. https://doi.org/10.1002/(SICI)1098-2329(199924)18:4<351::AID-ADV6>3.0.CO;2-X

[39]   P. Gatenholm, J. Kubat, A. Mathiasson. Biodegradable natural composites. I. Processing and properties, J. Appl. Polym. Sci., 45 (1992) 1667-1677. https://doi.org/10.1002/app.1992.070450918

[40]   S.M. Lee, D. Cho, W.H. Park, S.G. Lee, S.O. Han, L.T. Drzal. Novel silk/poly(butylene succinate) biocomposites: the effect of short fibre content on their mechanical and thermal properties. Compos. Sci. Technol., 65 (2005) 647-657. https://doi.org/10.1016/j.compscitech.2004.09.023

[41]   Z. Shao, F. Vollrath. Surprising strength of silkworm silk, Nature., 418 (2002) 741-741. https://doi.org/10.1038/418741a

[42]   T. Arai, G. Freddi, R. Innocenti, M. Tsukada. Preparation of water-repellent silks by a reaction with octadecenylsuccinic anhydride, J. Appl. Polym. Sci., 89 (2003) 324-332. https://doi.org/10.1002/app.12081

[43]   J. Summerscales, N.P.J. Dissanayake, A.S. Virk, W. Hall. A review of bast fibres and their composites. Part 1 – Fibres as reinforcements, Compos. A: Appl. Sci. Manuf., 41 (2010) 1329-1335. https://doi.org/10.1016/j.compositesa.2010.06.001

[44]   Y. Zou, H. Xu, Y. Yang. Lightweight polypropylene composites reinforced by long switchgrass stems, J. Polym. Environ., 18 (2010) 464-473. https://doi.org/10.1007/s10924-010-0165-4

[45]   W. Liu, A.K. Mohanty, L.T. Drzal, M. Misra. Novel biocomposites from native grass and soy based bioplastics: processing and properties evaluation, Indus. Eng. Chem. Res., 44 (2005) 7105–7112. https://doi.org/10.1021/ie050257b

[46]   M.R. Vignon, D. Dupeyre, C. Garcia-Jaldon. Morphological characterization of steam exploded hemp fibers and their utilization in propylene-based composites, Bioresource Biotechnol., 58 (1996) 203-215. https://doi.org/10.1016/S0960-8524(96)00100-9

[47]   M.M. Thwe, K. Liao. Effects of environmental aging on the mechanical properties of bamboo-glass fiber reinforced polymer matrix hybrid composites, Compos. Part A Appl. Sci. Manuf., 33 (2002) 43-52. https://doi.org/10.1016/S1359-835X(01)00071-9

[48]   Y. Lu, L. Weng, X. Cao. Morphological, thermal and mechanical properties of ramie crystallites—reinforced plasticized starch biocomposites, Carbohyd. Polym., 63 (2006) 198-204. https://doi.org/10.1016/j.carbpol.2005.08.027

[49]   E. Bodros, I. Pillin, N. Montrelay, C. Baley. Could biopolymers reinforced by randomly scattered flax fiber be used in structural applications? Compos. Sci. Technol., 67 (2007) 462-470. https://doi.org/10.1016/j.compscitech.2006.08.024

[50]   A.S. Singha, V.K. Thakur. Mechanical properties of natural fiber reinforced polymer composites, Bull. Mater. Sci., 31 (2008) 791-799. https://doi.org/10.1007/s12034-008-0126-x

[51]   H.H.C. Azeredo, L.H.C. Mattoso, D. Wood, T.G. Williams, R.J. Avena-Bustillos, T.H. Mchugh. Nanocomposite edible films from mango puree reinforced with cellulose nanofibers, J. Food Sci., 74 (2009) 31-35. https://doi.org/10.1111/j.1750-3841.2009.01186.x

[52] S. Ochi. Mechanical properties of Kenaf fibers and Kenaf/PLA composites, Mech. Mater., 40 (2008) 446-452. https://doi.org/10.1016/j.mechmat.2007.10.006

[53] N. Reddy. Extraction and characterization of natural cellulose fibers from common milkweed stems, Polym. Eng. Sci., 49 (2009) 2212-2217. https://doi.org/10.1002/pen.21469

[54] L. Famá, L. Gerschenson, S. Goyanes. Starch-vegetable fiber composites to protect food products, Carbohyd. Polym., 75 (2009) 230-235. https://doi.org/10.1016/j.carbpol.2008.06.018

[55] A. Kaushik, M. Singh, G. Verma. Green nanocomposites based on thermoplastic starch and steam exploded cellulose nanofibrils from wheat straw, Carbohyd. Polym., 82 (2010) 337-345. https://doi.org/10.1016/j.carbpol.2010.04.063

[56] P. Qu, Y. Gao, G. Wu, L. Zhang. Nanocomposites of poly (lactic acid) reinforced with cellulose nanofibrils. BioResources, 5 (2010) 1811-1823.

[57] D. Liu, T. Zhong, P.R. Chang, K. Li, Q. Wu. Starch composites reinforced by bamboo cellulosic crystals, Bioresource Technol., 101 (2010) 2529-2536. https://doi.org/10.1016/j.biortech.2009.11.058

[58] J.L. Guimarães, F. Wypych, C.K. Saul, L.P. Ramos, K.G. Satyanarayana. Studies of the processing and characterization of corn starch and its composites with banana and sugarcane fibers from Brazil, Carbohyd. Polym., 80 (2010) 130-138. https://doi.org/10.1016/j.carbpol.2009.11.002

[59] A. Ashori, A. Nourbakhsh. Bio-based composites from waste agricultural residues, Waste Manag., 30 (2010) 680-684. https://doi.org/10.1016/j.wasman.2009.08.003

[60] S.L. Favaro, M.S. Lopes, A.G. Vieira de Carvalho Neto, R. Rogerio de Santana, E. Radovanovic. Chemical, morphological, and mechanical analysis of rice husk/postconsumer polyethylene composites, Compos. A: Appl. Sci. Manuf., 41 (2010) 154-160. https://doi.org/10.1016/j.compositesa.2009.09.021

[61] D.P. Pfister, R.C. Larock. Green composites from a conjugated linseed oil-based resin and wheat straw, Compos. A: Appl. Sci. Manuf., 41 (2010) 1279-1288. https://doi.org/10.1016/j.compositesa.2010.05.012

[62] S. Ahankari, A.K. Mohanty, M. Misra. Mechanical behaviour of agro-residue reinforced poly(3-hydroxybutyrate-co-3-hydroxyvalerate), (PHBV) green composites: A comparison with traditional polypropylene composites, Compos. Sci.Technol., 71 (2011) 653- 657. https://doi.org/10.1016/j.compscitech.2011.01.007

[63]   N. Reddy, Y. Yang. Properties and potential applications of natural cellulose fibers from cornhusks, Green Chem., 7 (2005) 190-195. https://doi.org/10.1039/b415102j

[64]   A. Blicblau, R. Coutts, A. Sims. Novel composites utilizing raw wool and polyester resin, J. Mater. Sci. Lett., 16 (1997) 1417-1419. https://doi.org/10.1023/A:1018517512425

[65]   Y. Zhao, H.Y. Cheung, K.T. Lau, C.L. Lau, D.D. Zhao, H.L. Li. Silkworm silk/poly(lactic acid) biocomposites: Dynamic mechanical, thermal and biodegradable properties, Polym. Degrad. Stab., 95 (2010) 1978-1987. https://doi.org/10.1016/j.polymdegradstab.2010.07.015

[66]   C.K. Hong, R.P. Wool. Development of a bio-based composite material from soybean oil and keratin fibers, J. Appl. Polym. Sci.,95 (2005) 1524-1538. https://doi.org/10.1002/app.21044

[67]   A.J.F. Carvalho, A.A.S. Curvelo, J.A.M.A. Agnelli. First insight on composites of thermoplastic starch and kaolin, Carbohyd. Polym., 45 (2001) 189-194. https://doi.org/10.1016/S0144-8617(00)00315-5

[68]   J.K. Pandey, R.P. Singh. Green nanocomposites from renewable resources: Effect of plasticizer on the structure and material properties of clay-filled starch, Starch., 57 (2005) 8-15. https://doi.org/10.1002/star.200400313

[69]   J. Guan, M.A. Hanna. Selected morphological and functional properties of extruded acetylated starch-cellulose foams, Bioresource Technol., 97 (2006) 1716-1726. https://doi.org/10.1016/j.biortech.2004.09.017

[70]   A.P. Kumar, R.P. Singh. Biocomposites of cellulose reinforced starch: Improvement of properties by photo-induced crosslinking, Bioresource Technol., 99 (2008) 8803-8809. https://doi.org/10.1016/j.biortech.2008.04.045

[71]   X.F. Ma, J.G. Yu, N. Wang. Fly ash-reinforced thermoplastic starch composites, Carbohyd. Polym., 67 (2007) 32-39. https://doi.org/10.1016/j.carbpol.2006.04.012

[72]   A. Svagan. Bio-inspired cellulose Nanocomposites and foams based on starch matrix. PhD thesis, Department of Fiber and Polymer Technology, KTH Chemical Science and Engineering, SE-100 44, Stockholm, Sweden, 2008.

[73]   M. Avella, E. Martuscelli, B. Pascucci, M. Raimo, B. Focher, A. Marzetti. A new class of biodegradable materials: Poly-3-hydroxy-butyrate/steam exploded straw fiber composites, I. Thermal and impact behavior, J. Appl. Polym. Sci., 49 (1993) 2091-2103. https://doi.org/10.1002/app.1993.070491205

[74] B.S. Kaith, R. Jindal, A.K. Jana, M. Maiti. Development of corn starch based green composites reinforced with Saccharum spontaneum L fiber and graft copolymers—Evaluation of thermal, physico-chemical and mechanical propertie, Bioresource Technol., 101 (2010) 6843-6851. https://doi.org/10.1016/j.biortech.2010.03.113

[75] N. Ogata, G. Jimenez, H. Kawai, T. Ogihara. Structure and thermal/mechanical properties of poly(L-lactide)-clay blend, J. Polym. Sci. Part B: Polym. Phys., 35 (1997) 389-96. https://doi.org/10.1002/(SICI)1099-0488(19970130)35:2<389::AID-POLB14>3.0.CO;2-E

[76] R.S. Sinha, K. Okamoto, K. Yamada, M. Okamoto. Novel porous ceramic material via burning of polylactide/layered silicate nanocomposite, Nano Letts., 2 (2002) 423-426. https://doi.org/10.1021/nl020284g

[77] R.S. Sinha, K. Yamada, M. Okamoto, K. Ueda. New polylactide/layered silicate nanocomposite: A novel biodegradable material, Nano Letts., 2 (2002) 1093-1096. https://doi.org/10.1021/nl0202152

[78] R.S. Sinha, P. Maiti, M. Okamoto, K. Yamada, K. Ueda. New polylactide/layered silicate nanocomposites. 1. Preparation, characterization and properties., Macromolecule, 35 (2002) 3104-3110. https://doi.org/10.1021/ma011613e

[79] R.S. Sinha, K. Yamada, A. Ogami, M. Okamoto, K. Ueda. New polylactide layered silicate nanocomposite: Nanoscale control of multiple properties. Macromol, Rapid Commun., 23 (2002) 493-497.

[80] R.S. Sinha, M. Okamoto, K. Yamada, K. Ueda. New biodegradable polylactide/layered silicate nanocomposites: peparation, characterization and materials properties., Macromolecules, 35 (2002) 659-660.

[81] R.S. Sinha, M. Okamoto, K. Yamada, K. Ueda. New polylactide/layered silicate nanocomposites: Concurrent improvement of materials properties and biodegradability, Polymer., 44 (2003) 857-866. https://doi.org/10.1016/S0032-3861(02)00818-2

[82] K. Yamada, K. Ueda, R.S. Sinha, M. Okamoto. Preparation and properties of polylactide/layered silicate nanocomposites, Kobunshi Robunshu., 59 (2002) 760-765. https://doi.org/10.1295/koron.59.760

[83] P. Maiti, K. Yamada, M. Okamoto, K. Ueda, K. Okamoto. New polylactide/layered silicate nanocomposites: role of organoclay, Chem. Mater., 14 (2002) 4654-4661. https://doi.org/10.1021/cm020391b

[84] M.A. Paul, M. Alexandre, P. Degee, C. Calberg, R. Jerome, P. Dubois. Exfoliated polylactide/clay nanocomposites by in-situ coordination-insertion polymerization, Macromol., Rapid Commun., 24 (2003) 561-566. https://doi.org/10.1002/marc.200390082

[85] J.H. Lee, T.G. Park, H.S. Park, D.S. Lee, Y.K. Lee, S.C. Yoon, J.D. Nam. Thermal and mechanical characteristics of poly(L-lactic acid) nanocomposite scaffold, Biomaterials, 24 (2002) 2773-2778. https://doi.org/10.1016/S0142-9612(03)00080-2

[86] J. Chang, Y.U. An, D. Cho, E.P. Giannelis. Poly (lactic acid) nanocomposites: Comparison of their properties with montmorillonite and synthetic mica (II), Polymer, 44 (2003) 3715–3720. https://doi.org/10.1016/S0032-3861(03)00276-3

[87] D. Bondeson, K. Oksman. Dispersion and characteristics of surfactant modified cellulose whiskers nanocomposites, Compos. Interface, 14 (2007) 617-630. https://doi.org/10.1163/156855407782106519

[88] S. Lee, I. Kang, G. Doh, H. Yoon, B. Park, Q. Wu. Thermal and mechanical properties of wood flour/talc-filled polylactic acid composites: effect of filler content and coupling treatment, J. Thermoplast. Compos. Mater., 21 (2008) 209-223. https://doi.org/10.1177/0892705708089473

[89] M. Misra, H. Park, A.K. Mohanty, L.T. Drzal. Injection molded 'Green' nanocomposite materials from renewable resources. Presented at the Global Plastics Environmental Conference, Detroit, MI, USA, 18–19 February 2004.

[90] S.K. Mahadeva, S. Yun, J. Kim. Flexible humidity and temperature sensor based on cellulose-polypyrrole nanocomposite, Sensor. Actuator. A Phys., 165 (2011) 194-199.

[91] S. Tunç, O. Duman. Preparation of active antimicrobial methyl cellulose/carvacrol /montmorillonite nanocomposite films and investigation of carvacrol release, Food Sci. Technol., 44 (2011) 465-472. https://doi.org/10.1016/j.lwt.2010.08.018

[92] K.A. Zimmermann, J.M. LeBlanc, K.T. Sheets, R.W. Fox, P. Gatenholm. Biomimetic design of a bacterial cellulose/hydroxyapatite nanocomposite for bone healing applications, Mater. Sci. Eng., 31 (2011) 43-49. https://doi.org/10.1016/j.msec.2009.10.007

[93] S. Zadegan, M. Hosainalipour, H.R. Rezaie, H. Ghassai, M.A. Shokrgozar. Synthesis and biocompatibility evaluation of cellulose/hydroxyapatite

nanocomposite scaffold in 1-n-allyl-3-methylimidazolium chloride, Mater. Sci. Eng., 31 (2011) 954-961. https://doi.org/10.1016/j.msec.2011.02.021

[94] M.A. Sithique, M. Alagar. Preparation and properties of bio-based nanocomposites from epoxidized soy bean oil and layered silicate, Malaysian Polym. J., 5 (2010) 151-161.

[95] M. Wollerdorfer, H. Bader. Influence of natural fibres on the mechanical properties of biodegradable polymers, Indus. Crops Prod., 8 (1998) 105-112. https://doi.org/10.1016/S0926-6690(97)10015-2

[96] J.S. Tate, A.T. Akinola, D. Kabakov. Bio-based Nanocomposites: An Alternative to Traditional Composites, J. Technol. Stud., 1 (2010) 25-32.

[97] T.Y. Ke, X.Z. Sun. Effects of moisture content and heat treatment on the physical properties of starch and poly(lactic acid) blends, J. Appl. Polym. Sci., 81 (2001) 3069-82. https://doi.org/10.1002/app.1758

[98] T. Uesaka, K. Nakane, S. Maeda, T. Ogihara, N. Ogata. Structure and physical properties of poly(butylene succinate)/cellulose acetate blends, Polymer, 41 (2000) 8449-8454. https://doi.org/10.1016/S0032-3861(00)00206-8

[99] C.D. Kesel, C.V. Wauven, C. David. Biodegradation of polycaprolactone and its blends with poly(vinylalcohol) by micro-organisms from a compost of house-hold refuse, Polym. Degrad. Stab., 55 (1997) 107-113. https://doi.org/10.1016/0141-3910(95)00138-7

[100] L. Averous, N. Fauconnier, L. Moro. Fringant blends of thermoplastic starch and polyesteramide: processing and properties, J. Appl. Polym. Sci., 76 (200) 1117-1128.

[101] J.L. Willett, R.L. Shogren. Processing and properties of extruded starch/polymer foams, Polymer, 43 (2002) 5935-5947. https://doi.org/10.1016/S0032-3861(02)00497-4

[102] O. Martin, L. Averous. Poly (lactic acid): Plasticization and properties of biodegradable multiphase systems, Polymer, 42 (2001) 6209-6219. https://doi.org/10.1016/S0032-3861(01)00086-6

[103] P. Sarazin, G. Li, W.J. Orts, B.D. Favis. Binary and ternary blends of polylactide, polycaprolactone and thermoplastic starch, Polymer, 49 (2008) 599- 609. https://doi.org/10.1016/j.polymer.2007.11.029

[104] K. Majdzadeh-Ardakani, Sh. Sadeghi-Ardakani. Experimental investigation of mechanical properties of Starch/natural rubber/clay nanocomposites, Digest J. Nanomater. Biostruct, 5 (2010) 307-316.

[105] P. Maiti, C.A. Batt, E.P. Giannelis. Renewable plastics: Synthesis and properties of PHB nanocomposites, Polym. Mater. Sci. Eng., 88 (2003) 58-59.

[106] J.P. Zheng, P. Li, Y.L. Ma, K.D. Yao. Gelatine/montmorillonite hybrid nanocomposite - preparation and properties, J. Appl. Polym. Sci., 86 (2002) 1189-1194. https://doi.org/10.1002/app.11062

[107] A. Takegawa, M. Murakami, Y. Kaneko, J. Kadokawa. Preparation of chitin/cellulose composite gels and films with ionic liquids, Carbohyd. Polym., 79 (2010) 85-90. https://doi.org/10.1016/j.carbpol.2009.07.030

[108] M.R.S. Nunes, R.C. Silva, Jr. J.G. Silva, J. Tonholo, A.S. Ribeiro. Preparation and morphological characterization of chitosan/clay nanocomposites. In Proceedings of the 11th International Conference on Advanced Materials, Rio de jenero, Brazil, 20–25 September (2009) 20-25.

[109] R. Talreja, J.A.E. Ma°nson. Polymer Matrix Composites. Elsevier Science, Amsterdam, Netherlands, (2001) 403-432.

[110] P.A. Fowler, J.M. Hughes, R.M. Elias, Biocomposites: technology, environmental credentials and market forces, J. Sci. Food. Agric., 86 (2006) 1781-1789. https://doi.org/10.1002/jsfa.2558

[111] A.K. Bledzki, J. Gassan. Composites reinforced with cellulose based fibers, Prog. Polym. Sci., 24 (1999) 221-274. https://doi.org/10.1016/S0079-6700(98)00018-5

[112] A. Arbelaiz, B. Fernandez, J.A. Ramos, I. Mondragon. Thermal and crystallization studies of short flax fibre reinforced polypropylene matrix composites: Effect of treatments, Thermochim. Acta, 440 (2006) 111-121. https://doi.org/10.1016/j.tca.2005.10.016

[113] M.N. Belgacem, A. Gandini. The surface modification of cellulose fibres for use as reinforcing elements in composite materials, Compos. Interf., 12 (2005) 41-75. https://doi.org/10.1163/1568554053542188

[114] J. George, M.S. Sreekala, S. Thomas. A review on interface modification and characterization of natural fiber reinforced plastic composites., Polym. Eng. Sci., 41 (2001) 1471-1485. https://doi.org/10.1002/pen.10846

[115] S. Kalia, B.S. Kaith, I. Kaur. Pretreatments of natural fibers and their application as reinforcing material in polymer composites—A review, Polym. Eng. Sci., 49 (2009) 1253-1272. https://doi.org/10.1002/pen.21328

[116] D. Maldas, B.V. Kokta, R.G. Raj, C. Daneault. Improvement of the mechanical properties of sawdust wood fibre—polystyrene composites by chemical treatment, Polymer, 29 (1988) 1255-1265. https://doi.org/10.1016/0032-3861(88)90053-5

[117] M. Baiardo, G. Frisoni, M. Scandola, A. Licciardello. Surface chemical modification of natural cellulose fibers, J. Appl. Polym. Sci., 83 (2002) 38-45. https://doi.org/10.1002/app.2229

[118] M. Baiardo, E. Zini, M. Scandola. Flax fibre-polyester composites, Compos. A: Appl. Sci. Manuf., 35 (2004) 703-710. https://doi.org/10.1016/j.compositesa.2004.02.004

[119] G. Frisoni, M. Baiardo, M. Scandola, D. Lednicka, M.C Cnockaert, J. Mergaert, J. Swings. Natural Cellulose Fibers: Heterogeneous acetylation kinetics and biodegradation behavior, Biomacromolecules, 2 (2001) 476-482. https://doi.org/10.1021/bm0056409

[120] E. Zini, M. Scandola, P. Gatenholm. Heterogeneous Acylation of Flax Fibers. Reaction Kinetics and Surface Properties, Biomacromolecules, 4 (2003) 821-827. https://doi.org/10.1021/bm034040h

[121] E. Zini, M. Baiardo, M. Scandola. Biodegradable polyesters reinforced with surface-modified vegetable fibers, Macromol. Biosci., 4 (2004) 286-295. https://doi.org/10.1002/mabi.200300120

[122] E. Zini, M.L. Focarete, I. Noda, M. Scandola. Bio-composite of bacterial poly(3-hydroxybutyrate-co-3-hydroxyhexanoate) reinforced with vegetable fibers, Compos. Sci. Technol., 67 (2007) 2085-2094. https://doi.org/10.1016/j.compscitech.2006.11.015

[123] P. Tran, D. Graiver, R. Narayan. Biocomposites synthesized from chemically modified soy oil and biofibers, J. Appl. Polym. Sci., 102 (2006) 69-75. https://doi.org/10.1002/app.22265

[124] N. Lee, O.J. Kwon, B. Chun, J. Cho, J.S. Park. Characterization of castor oil/polycaprolactone polyurethane biocomposites reinforced with hemp fibers, Fibers Polym., 10 (2009) 154-160. https://doi.org/10.1007/s12221-009-0154-1

[125] K.S. Thomas, C. Pavithran. Effect of chemical treatment on the tensile properties of short sisal fibre-reinforced polyethylene composites, Polymer, 37 (1996) 5139-5149. https://doi.org/10.1016/0032-3861(96)00144-9

[126] N.M. Belgacem, P. Bataille, S. Sapieha. Effect of corona modification on the mechanical properties of polypropylene/cellulose composites, J. Appl. Polym., Sci. 53 (1994) 379-385. https://doi.org/10.1002/app.1994.070530401

[127] Z.F. Li, A.N. Netravali. Surface modification of UHSPE fibres through allylamine plasma deposition II. Effect on fibre and fibre/epoxy interface, J. Appl. Polym. Sci., 44 (1992) 333-346. https://doi.org/10.1002/app.1992.070440217

[128] J. Gassan, V.S. Gutowski. Effect of corona discharge and UV treatment on the properties of jute–fiber epoxy composites, Compos. Sci. Technol., 60 (2000) 2857-2863. https://doi.org/10.1016/S0266-3538(00)00168-8

[129] C.S. Wu. Renewable resource-based composites of recycled natural fibers and maleated polylactide bioplastic: Characterization and biodegradability, Polym. Degrad. Stab., 94 (2009) 1076-1084. https://doi.org/10.1016/j.polymdegradstab.2009.04.002

[130] R. Karnani, M. Krishnan, R. Narayan. Biofiber-reinforced polypropylene composites, Polym. Eng. Sci., 37 (1997) 476-483. https://doi.org/10.1002/pen.11691

[131] G. Canche´-Escamilla, J.I. Cauich-Cupul, E. Mendizabal, H. Vazquez-Torres, P.J. Herrera-Franco, Mechanical properties of acrylate-grafted henequen cellulose fibers and their application in composites, Compos A., 30 (1999) 349-359. https://doi.org/10.1016/S1359-835X(98)00116-X

[132] A.L. Martinez-Hernandez, C. Velasco-Santos, M.D. Icaza, V.M. Castano. Grafting of methyl methacrylate onto natural keratin, e-Polymers, 16 (2003) 1-11.

[133] S. Wong, R.A. Shanks, A. Hodzic. Effect of additives on the interfacial strength of poly (l-lactic acid) and poly (3-hydroxy butyric acid)-flax fibre composites, Compos. Sci. Technol., 67 (2007) 2478-2484. https://doi.org/10.1016/j.compscitech.2006.12.016

[134] S.N. Khot, J.J. Lascala, E. Can, S.S. Morye, G.I. Williams, G.R. Palmese, S.H. Kusefoglu, R.P. Wool. Development and application of triglyceride-based polymers and composites, J. Appl. Polym. Sci., 82 (2001) 703-723. https://doi.org/10.1002/app.1897

[135] B.N. Melo, C.G. dos-Santos, V.R. Botaro, V.M.D. Pasa. Eco-composites of polyurethane and Luffa aegyptiaca modified by mercerisation and benzylation, Polym. Polym. Compos., 16 (2008) 249-256.

[136] K. Adekunle, D. Akesson, M. Skrifvars. Biobased composites prepared by compression molding with a novel thermoset resin from soybean oil and a natural fiber reinforcement, J. Appl. Polym. Sci., 116 (2010) 1759-1765. https://doi.org/10.1002/app.31634

[137] S. Dutta, N. Karak, S. Baruah. Jute-fiber-reinforced polyurethane green composites based on Mesua ferrea L. seed oil, J. Appl. Polym. Sci., 115 (2010) 843-850. https://doi.org/10.1002/app.30357

[138] N. Boquillon. Use of an epoxidized oil-based resin as matrix in vegetable fibers-reinforced composites, J. Appl. Polym. Sci., 101 (2006) 4037-4043. https://doi.org/10.1002/app.23133

[139] Z. Liu, S.Z. Erhan, D.E. Akin, F.E. Barton. "Green" composites from renewable resources: preparation of epoxidized soybean oil and flax fiber composites, J. Agric. Food Chem., 54 (2006) 2134-2137. https://doi.org/10.1021/jf0526745

[140] R.V. Silva, D. Spinelli, W.W.B. Filho, S.C. Neto, G.O. Chierice, J.R. Tarpani. Fracture toughness of natural fibers/castor oil polyurethane composites, Compos. Sci. Technol., 66 (2006) 1328-1335. https://doi.org/10.1016/j.compscitech.2005.10.012

[141] G.I. Williams, R.P. Wool. Composites from natural fibers and soy oil resins, Appl. Compos. Mater., 7 (2000) 421-432. https://doi.org/10.1023/A:1026583404899

[142] N. Graupner, A.S. Herrmann, J.M. Ssig. Natural and man-made cellulose fibre-reinforced poly(lactic acid) (PLA) composites: an overview about mechanical characteristics and application areas, Compos. A: Appl. Sci. Manuf., 40 (2009) 810-821. https://doi.org/10.1016/j.compositesa.2009.04.003

[143] E.R. Coats, F.J. Loge, M.P. Wolcott, K. Englund, A.G. McDonald. Production of natural fiber reinforced thermoplastic composites through the use of polyhydroxybutyrate-rich biomass, Bioresour. Technol., 99 (2008) 2680-2686. https://doi.org/10.1016/j.biortech.2007.03.065

[144] S. Luo, A.N. Netravali. Mechanical and thermal properties of environment-friendly "green" composites made from pineapple leaf fibers and poly (hydroxybutyrate-co-valerate) resin, Polym. Compos., 20 (1999) 367-378. https://doi.org/10.1002/pc.10363

[145] S. Singh, A.K. Mohanty. Wood fiber reinforced bacterial bioplastic composites: Fabrication and performance evaluation, Compos. Sci. Technol., 67 (2007) 1753-1763. https://doi.org/10.1016/j.compscitech.2006.11.009

[146] L. Averous, N. Boquillon. Biocomposites based on plasticized starch: thermal and mechanical behaviours, Carbohydr. Polym., 56 (2004) 111-122. https://doi.org/10.1016/j.carbpol.2003.11.015

[147] M. Morreale, R. Scaffaro, A. Maio, F.P. La Mantia. Effect of adding wood flour to the physical properties of a biodegradable polymer, Compos. A: Appl. Sci. Manuf., 39 (2008) 503-513. https://doi.org/10.1016/j.compositesa.2007.12.002

[148] N.M. Barkoula, S.K. Garkhail, T. Peijs. Biodegradable composites based on flax/polyhydroxybutyrate and its copolymer with hydroxyvalerate, Indus. Crops Prod., 31 (2010) 34-42. https://doi.org/10.1016/j.indcrop.2009.08.005

[149] A.K. Mohanty, A. Wibowo, L.T. Drzal. Effect of process engineering on the performance of natural fiber reinforced cellulose acetate biocomposites, Compos. A: Appl. Sci. Manuf., 35 (2004) 363-370. https://doi.org/10.1016/j.compositesa.2003.09.015

[150] A. Aluigi, C. Vineis, A. Ceria, C. Tonin. Composite biomaterials from fibre wastes: Characterization of wool-cellulose acetate blends, Compos. A: Appl. Sci. Manuf., 39 (2008) 126-132. https://doi.org/10.1016/j.compositesa.2007.08.022

[151] M. Carus, C. Gahle. Injection moulding with natural fibres, Reinf. Plast., 52 (2008) 18-25. https://doi.org/10.1016/S0034-3617(08)70101-2

[152] European Commission, L269, Official Journal of European Commission, 21st October 2000 (Directive 2000/53/EC), Life Focus (2004).

[153] L.A. Duigou, P. Davies, C. Baley. Seawater ageing of flax/poly(lactic acid) biocomposites, Polym. Degrad. Stab., 94 (2009) 1151-1162. https://doi.org/10.1016/j.polymdegradstab.2009.03.025

Chapter 2

Polyaniline (PANI) Based Composites for the Adsorptive Treatment of Polluted Water

Abu Nasar

Department of Applied Chemistry, Faculty of Engineering & Technology, Aligarh Muslim University, Aligarh – 202002, India

abunasaramu@gmail.com

Abstract

Adsorption techniques are commonly applied for the treatment of pollutants from wastewater. Different varieties of adsorbents have been used for removing contaminants from aqueous media. Although conventional adsorbents (like activated carbon, alumina, etc.) should be preferable, their extensive application is undesirable because of the high cost and regeneration problems. To overcome these problems, attention has been directed to developing cheaper and efficient alternatives to conventional adsorbents. In this respect, the polyaniline (PANI) based composites offer suitable substitutes for conventional adsorbents due to their special properties such as easy preparation, nontoxicity and availability as active functional groups which have interactions with contaminating molecules and ions. In the present chapter different adsorption parameters, isotherms and kinetics along with their significance are described. Detailed discussions have been dedicated to the applications of PANI-based composites for the adsorptive elimination of dyes and toxic metals. The role of these adsorbents in monitoring of pesticides in water has also been discussed briefly.

Keywords

Wastewater Treatment, PANI-based Composites, Dyes, Heavy Metals, Pesticides

Contents

1.   Introduction

One of the major difficulties of the present time is that the world is observing pollution which is growing with time and producing severe harm to human health. The pollution is any objectionable alteration in any physical or chemical properties of air, soil and water that makes the environment unnatural and creates health hazards to the living organisms. Water is a precious natural resource without which life is impossible. The adulteration of water resources by different categories of pollutants is a matter of great worry. The pollution of water resources (rivers, lakes, groundwater, etc.) occurs due to the discharge of untreated wastewater released from domestic, commercial and industrial activities.

Water used in the industry generates a wastewater that has possible hazards for human because it introduces different toxic impurities such as dyes and heavy metals into water resources. A large number of biological, inorganic and organic impurities have been reported as water contaminants [1]. The presence of dyes, pesticides and heavy metals plays a major role in the contamination of water. These contaminants have different negativity and come from different sources. The synthetic dyes are commonly used by various industries like textiles, food, cosmetics, leather, plastic, glass, medicine, ceramics, pharmaceuticals, printing ink, paper and pulp, etc. These industries are using large amounts of water in their manufacturing processes and thereby releasing large amounts of dye-loaded effluents into the freshwater systems. These dyes are known to have long-term damaging effects. Dyes disturb the photosynthetic activities of the aquatic flora by absorbing and reflecting the sunlight entering the water bodies and thus severely affecting the food chain. Most of the synthetic dyes are dangerous to living organisms [2-6]. Even in very small quantities, dyes are visible to the naked eye and introduce a disorder to the ecosystem and water bodies [6,7]. Thus, the treatment of dyes prior to discharge of contaminated water is a major challenge for the environmentalists. Like dyes, pesticides are also prominent contaminants of water resources. Further, to fulfill the growing demand of crop production pesticides have been used throughout the world. The extensive use of the various pesticides attracted increasing researcher attention due to their existence in food products, soils and water resources [8-15].

Further, as far as heavy metals are concerned, they cannot be degraded into simpler forms, and therefore they enter the inside of living beings as inorganic salts or organometallic derivatives. Many metals are considered as water pollutants with a high degree of toxicity and other harmful effects [16-20]. In fact, the ground water pollution due to industrial wastewater by dyes and heavy metals ions has become a universal environmental problem in several countries. Thus the treatment of industrial wastewater is essential before its discharge. Various treatment technologies have been stated in the literature for the elimination of dyes from wastewater like membrane filtration, ozonation, Fenton's process, reverse osmosis, coagulation, etc. However, amongst the various available methods, adsorption technique was established to be one of the most effective and superior techniques because of the various advantages associated with it i.e. low cost, insensitivity to toxic pollutants, simplicity and flexibility of design [3–5,7,16,17].

2. Adsorption phenomenon

It is a process of adhesion of atoms, molecules or ions from a gas, liquid or solution on the surface of another substance. When a solid surface is exposed to gas/liquid/solution, the atoms/ions/molecules from the later phase accumulate or concentrate at the surface of the former. This phenomenon of concentration or accumulation of species on a solid surface is called as adsorption. The adsorption phenomenon is associated with the involvement of two components, namely adsorbate and adsorbent. The adsorbent is the material which offers the surface on which adsorption occurs while the adsorbate is the one which is being accumulated on the surface of adsorbent i.e. adsorbate becomes attached on the surface of the adsorbent (Fig. 1).

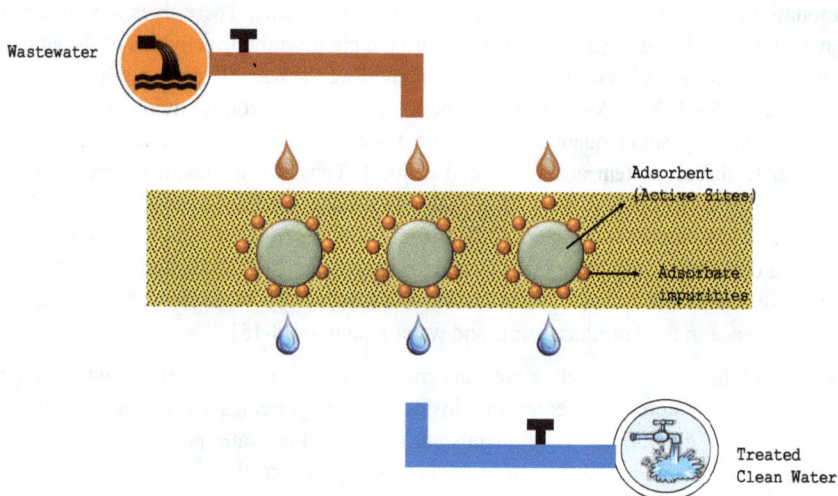

Fig. 1 Treatment of impurities by adsorption.

In bulk of solid adsorbent, all the bonding necessities of the component atoms are fulfilled by other atoms while surface atoms are not entirely enclosed by other atoms or molecules and there exists a residual force. Because of the residual force, the adsorbent surface attracts adsorbate. Depending upon the nature of the attracting force the adsorption phenomenon is categorized as physisorption (due to weak Van Der Waals forces) or chemisorption (due to covalent bonding).

Adsorption is a well-known effective technique used for the handling of wastewater released from domestic and industrial activities. In the wastewater treatment, the adsorbate is polluted water from which the solute (contaminants, e.g., dyes, heavy metals etc.) is attached to adsorbent which is a porous solid material (having high surface area). With the progress of adsorption process, the equilibrium between the adsorbent and solution is achieved. The extent of adsorption is represented by two, namely, removal efficiency (also known as percent removal, % R) and adsorption capacity (q_e). The quantity of contaminant adsorbed at equilibrium can be obtained by using the equations:

$$q_e = \frac{C_o - C_e}{m} V$$

$$\% R = \frac{C_o - C_e}{C_o} \times 100$$

where, m is the mass (g) of the adsorbent, V is the volume of solution (L), and C_o and C_e are the initial and equilibrium concentrations (mg/L) of adsorbate, respectively.

2.1 Adsorption isotherm

The amount of adsorbate adsorbed per unit weight of adsorbent as a function of the equilibrium concentration at a fixed temperature, known as adsorption isotherm, is used to explain the nature of the adsorbate-adsorbent system. Langmuir, Freundlich, Temkin, Dubinin-Radushkevich, Sips adsorption isotherm models are generally used to represent the adsorption data.

2.1.1 Langmuir adsorption isotherm

This isotherm model [21] is most popular and very commonly used to represent a variety of adsorbate-adsorbent systems. This model assumes monolayer coverage of adsorbate on the surface of adsorbent and mathematically represented respectively in both non-linear and linear form:

$$q_e = \frac{q_m K_L C_e}{1 + K_L C_e}$$

$$\frac{1}{q_e} = \frac{1}{q_m K_L C_e} + \frac{1}{q_m}$$

where, q_m is the maximum adsorption capacity (mg/g) for the formation of a complete monolayer on the adsorbent surface and K_L (L/mg) is Langmuir constant associated to the affinity between the adsorbate and adsorbent. Thus a linear variation of a graphical plot of $1/q_e$ against $1/C_e$ shows the validity of Langmuir adsorption isotherm in a given adsorbate-adsorbent system (Fig. 2a). This is an indication of monolayer coverage of dye molecules and a chemisorption type of attachment between the adsorbate and adsorbent. Langmuir constants q_m and K_L can be determined from the slope and interception of the plot.

2.1.2 Freundlich adsorption isotherm

This isotherm is an empirical model is based on the assumption that adsorption occurs over heterogeneous adsorption surface having the adsorption sites of different energies and are not equally available [22]. This isotherm can be mathematically represented respectively in the non-linear and linear forms as:

$$q_e = K_F C_e^{\frac{1}{n}}$$

$$\ln q_e = \frac{1}{n}\ln C_e + \ln K_F$$

where, K_F ($mg^{1-1/n}L^{1/n}/g$) and n are Freundlich constants related to the adsorption capacity and adsorption intensity, respectively. The validity of this isotherm can be checked by plotting $\ln q_e$ against $\ln C_e$ (Fig. 2b).

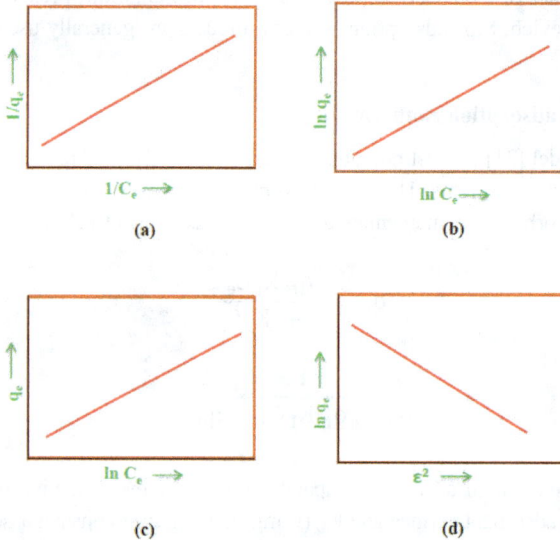

(a)

(b)

(c)

(d)

Fig. 2 Symbolic plots of different isotherm models: (a) Langmuir Adsorption Isotherm, (b) Freundlich Adsorption Isotherm, (c) Temkin adsorption isotherm, (d) Dubinin–Radushkevich (DR) adsorption isotherm

2.1.3 Temkin adsorption isotherm

The Temkin model is based on the assumption that the enthalpy of adsorption on the surface decreases linearly with the coverage [23]. This isotherm in its mathematical linear form is represented as:

$$q_e = B \ln C_e + B \ln K_T$$

where, $B = RT/b$, b is Temkin constant related with the enthalpy of adsorption (J/mol), R is gas constant (J/K/mol) and K_T is the equilibrium binding constant (L/g). The isotherm is best represented by a linear q_e vs. $\ln C_e$ plot (Fig. 2c).

2.1.4 Dubinin–Radushkevich (DR) adsorption isotherm

The mathematical form of Dubinin-Radushkevich (D-R) isotherm model [24] is represented as follows:

$$\ln q_e = \ln q_D - \beta \varepsilon^2$$

$$\varepsilon = RT \ln(1 + 1/C_e)$$

where, ε is the polyani potential, β is a constant related to the mean adsorption energy (E) which is given as below:

$$E = \frac{1}{\sqrt{2\beta}}$$

2.1.5 Suitability of isotherm model

As far as the best suitability of adsorption model, it is relevant to mention here that the different isotherms are applied to the adsorption for case by case. If a model is suitable for a specific adsorbent–adsorbate system the same may not be valid for other systems. For example, Kanwal et. al [25], while studying the removal of Cr (III), reported that Freundlich model obeyed better than Langmuir when polyaniline (PANI) was used as an adsorbent while Langmuir isotherm model holds good with PANI/Sawdust and PANI/Rice husk composites. In fact, the suitability of the best isotherm can be best judged by analyzing the correlation coefficient for the different linear plots of different isotherm models.

2.2 Adsorption kinetics

The kinetic studies are a fundamental approach for monitoring and understanding the behavior of any adsorbate-adsorbent system. There are a number of models such as pseudo-first order, pseudo-second order, Elovich, intraparticle diffusion etc. which are available to analyze the kinetics behind the adsorption phenomenon. The mathematical form of pseudo-first order, pseudo-second order, Elovich and intraparticle diffusion models can be respectively expressed as [4]:

$$\ln(q_e - q_t) = \ln q_e - k_1 t$$

$$\frac{t}{q_t} = \frac{t}{q_e} + \frac{1}{k_2 q_e^2}$$

$$q_t = \frac{1}{\beta}\ln(\alpha\beta) + \frac{1}{\beta}\ln t$$

$$q_t = k_{id} t^{1/2} + C$$

where, k_1 is the rate constant (min^{-1}) of pseudo-first order reaction, k_2 is pseudo-second order rate constant (g mg^{-1} min^{-1}), K_{id} is intraparticle diffusion rate constant (mg g^{-1} min$^{-1/2}$), C is the thickness of the boundary layer, α is the initial rate of adsorption (mg g^{-1} min^{-1}) and β is related to surface coverage (g mg^{-1}), q_t is the adsorption capacity (mg g^{-1}) at time t and q_e is the adsorption capacity (mg g^{-1}) at equilibrium. The qualitative aspect these four models has been represented by symbolic graphical plots in Fig. 3.

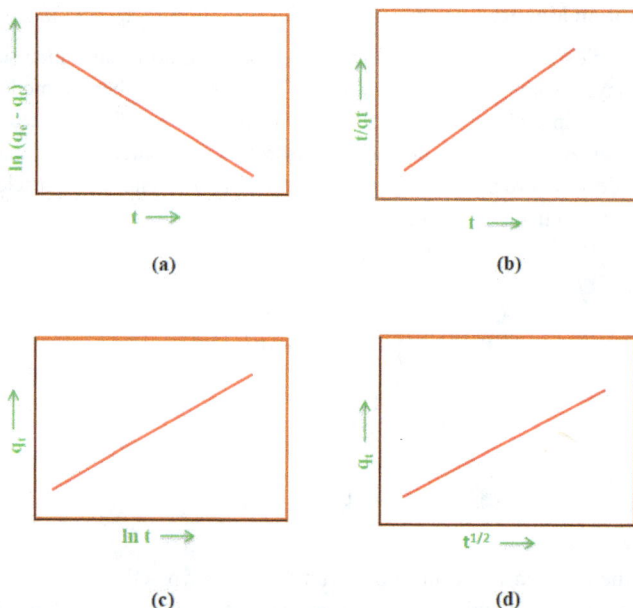

Fig. 3 Symbolic representation of different kinetic models: (a) pseudo-first order, (b) pseudo-second order, (c) intraparticle diffusion and (d) Elovich plot.

3. Types of adsorbents

A good adsorbent should a have large surface area, requires less time to achieve equilibrium and have large adsorption capacity. Both conventional and non-conventional adsorbents have been used for the treatment of wastewater. Conventional adsorbents include activated carbon, activated alumina, silica gel and activated clay while non-conventional adsorbents include inorganic, organic and biosorbent materials of different categories. Among the variety of adsorbents, activated carbon would logically be the most preferred adsorbent for the removal of contaminants. Because of the high surface area, microporous structure and outstanding adsorption power, activated carbon is undoubtedly considered as a universal adsorbents, however, its widespread use is sometimes restricted due to the high cost and difficulty in regeneration. Thus, devotion has been directed towards the preparation of new and cheaper substitutes of activated carbon and other conventional adsorbents. A variety of non-conventional cheaper adsorbents obtained from domestic, agricultural, industrial wastes such as coir pith, orange peel, banana peel, rice husk, citrus limetta peel, date pit, jackfruit peel, garlic, oil palm trunk, guava leaf powder, almond shell, hazelnut shell, pineapple stem, peanut hull, punica granatum peel, wheat straw, egg shell, longan shell, spent tea leaves corn cobs, used tea leaves, metal hydroxide sludge, red mud, fly ash, bentonite, diatomite, zeolites, chitosan, peat, biomass, starch, cotton, etc. have been used for the treatment of different types of polluted wastewater. Further, during recent years there has been an increasing trend to improve the adsorption efficiency by chemically modifying these low-cost adsorbents. It has generally been observed that chemically modified adsorbents have higher adsorption capacities than their unmodified forms. In this context, the polyaniline (PANI) based composites appeared to be a potential substitute because of their simplicity of preparation, porosity and water insolubility.

4. Polyaniline (PANI) based composites as adsorbents

Polyaniline (PANI) is a renowned conducting polymer and advantageous owing to its affordable price, easy synthesis, good conductivity, mechanical flexibility, environmental stability, etc. The potential use of PANI, as an adsorbent, in wastewater treatment is due to the presence of active groups, namely amine and imine, which have interactions with molecules of various contaminants present in polluted water [26]. Both inorganic materials (e.g. silica, montmorillonite, allapulgite, Fe_3O_4, $ZnCl_2$, $CuCl_2$ etc.) as well as organic materials (e.g. sawdust, polysaccharides, agricultural wastes, etc.) have been used to prepare PANI-based composites as adsorbents for the effective removal of water contaminants.

4.1 PANI-based adsorbents for the treatment of dye-contaminated wastewater

Ansari and Mosayebzadeh [27] used PANI coated sawdust for the treatment of methyl orange dye from water and the removal efficiency was found to be higher compared to activated carbon. The correlation coefficient obtained for the Langmuir model was greater than that for the Freundlich model, signifying a higher possibility of monolayer adsorption over multilayer coverage. The kinetics of adsorption was best obeyed by pseudo-second order. Potash alum doped PANI has successfully been used for the selective treatment of a number of sulfonated anionic dyes from water [28]. Patil and Shrivastava [29] have synthesized PANI-Fe_2O_3 magnetic nanocomposite from Fe_2O_3 nano-particles, monomer aniline and ammonium persulfate and successfully utilized this composite for the effective removal acid violet 19 dye from aqueous solution. In a very recent work [30] waterborne polyaniline (PANI) core–shell nanoparticles were synthesized via chemical oxidation of aniline in presence of polyvinyl pyrrolidone (PVP) as a stabilizer in acidic medium. This waterborne polyvinyl pyrrolidone (PVP) stabilized polyaniline (PANI) core–shell nanoparticles have been used for the removal of methyl orange dye from aqueous solutions. The result was best represented by Langmuir isotherm indicating the uniform monolayer coverage of dye adsorbate on the adsorbent surface. It was also reported that the interaction of anionic methyl orange with the adsorbent was the electrostatic and highly selective presence of other anions, e.g. SO_4^{-2}, NO_3^- and halogens.

Laabd et al [31] recently synthesized PANI/Bi_2WO_6 (Polyaniline/Bismuth tungstate) nanocomposites by oxidation polymerization of aniline in the presence of Bi_2WO_6 nanoparticles in an acidic solution. This bismuth tungstate co-modified by PANI was successfully utilized to eliminate anionic Congo red (CR) from the water. Their studies showed the CR dye removal efficiency is significantly influenced by different parameters such as initial pH, contact time, PANI weight content in nano-composite, adsorbent dose, CR initial concentration and temperature. They also reported that CR adsorption efficiency increased from 13.7 to 92.0% when the PANI content coated on Bi_2WO_6 nanoparticles was increased from 0 to 10 wt.%, respectively. Further, the experimental results were observed to be best followed by the pseudo-second order kinetic while isotherm data were best fitted by the Redlich-Peterson and Langmuir isotherm models.

In an early work [32], methylene blue was removed by a nanocomposite made of PANI coated wood sawdust (PANI/SD). This composite was prepared via direct chemical polymerization onto sawdust which was initially soaked in aniline in acidic (HCl) media. It was submitted that this adsorbent was eco-friendly, cheap and regenerable and accordingly it may be considered as an appropriate adsorbent for the treatment of textile wastewater contaminated by methylene blue. Methylene blue from water has also been recently removed by polyaniline-nickel ferrite (PANI/$NiFe_2O_4$) nano-composite [33]. An eco-friendly polymer composite, Polyaniline/chitosan, was used as an adsorbent for the treatment of a number of dyes and the maximum percentage removal were found to be 95.4, 98.2% and 99.8% for the Congo red, coomassie, brilliant blue, and remazol brilliant blue R, respectively [34]. Langmuir isotherm and pseudo-second-order kinetic models were reported to be followed by the removal all the above dyes. While studying the treatment of methylene blue dye from aqueous solution by using the adsorbent made of polyaniline nanotubes (PANI-NTs) base/silica composite as adsorbent Ayad et al [35] found that the efficiency of this composite is much better than the conventional PANI based/silica composite powder for dye uptake. Again in this case also the isotherm was best obeyed by the Langmuir model and kinetics by pseudo-second order model. In similar way the adsorptive removal of Rhodamine B and Congo red by PANI/α-MoO_3

[36], methylene blue by polyaniline nanotubes base/silica [37] and orange 16 by chitosan–polyaniline/ZnO hybrid [38] composites have been observed to be well fitted by Langmuir isotherm along with the pseudo-second-order kinetics. However, the adsorptive elimination of methylene blue by polyaniline/polyvinyl alcohol/clinoptilolite nanocomposite, as prepared by oxidation polymerization of anilinium cations inside and outside of the clinoptilolite channels, was reported to follow the intraparticle diffusion kinetic model [39]. In most of the reported work, it has been reported that the adsorptive removal of dyes by PANI-based composites generally obey the Langmuir adsorption isotherm model i.e. associated with the monolayer film formation on the adsorbent surface. However, Haitham et al [40] obtained a different trend on the removal of methyl orange by a recyclable immobilized polyaniline on a glass plate. They reported that the suspended powder system obeys the Langmuir model better than the Freundlich isotherm while the immobilized system obeys Freundlich better than the Langmuir model. This suggests that surface heterogeneity is more pronounced when the adsorbent is immobilized on a flat surface. Further, in the treatment of tartrazine dye from water by PANI/Fe$_2$O$_3$ has also been observed to follow Freundlich adsorption isotherm [41]. The cited literature (Table 1) indicates that kinetic behavior of most of the work on the applications of PANI-based composites for dye removal is best represented by pseudo-second kinetics. However, recently Rafiqi and Majid [42], on the basis of removal study of methylene blue dye from water by polyaniline and polyaniline–nitroprusside (PANI/NP) composite, reported that the pseudo-first order is the most appropriate kinetic model to represent the adsorption of methylene blue onto polyaniline and pseudo-second order onto PANI–NP composite.

4.2   PANI-based adsorbents for the monitoring/treatment of pesticides contaminated wastewater

A number of different methods such as photocatalytic degradation [43,44], advanced oxidation processes [45], nanofiltration membranes [46], aerobic degradation [47], solid phase extraction [48], fluid extraction [49], colloidal manganese dioxide oxidation [50-54], combined photo-Fenton and biological oxidation [55], adsorption [56-61] etc. have been used for the treatment of pesticides. Although, so far, no considerable work on the adsorptive treatment of pesticide by using PANI-based composite as adsorbent has been performed some extensive investigations have been carried out for monitoring and testing the pesticide by using this composite. The superparamagnetic attapulgite/Fe$_3$O$_4$/polyaniline (ATP/Fe$_3$O$_4$/PANI) nanocomposite was successfully employed as a sorbent for magnetic dispersive solid phase extraction of benzoylurea insecticides in water samples and was reported to be very effective in monitoring the insecticide in river samples. This ATP/Fe$_3$O$_4$/PANI nanocomposite was prepared by a one-pot method, which involved the applications of Fe(III) as both the oxidant for the polymerization of aniline and the single iron source of Fe_3O_4 formed by the redox reaction between aniline and Fe(III). Mehdinia [63] has prepared the chemically improved filter paper with polyaniline/polypyrrole cellulose nanocomposite and utilised as the solid phase sorbent for preconcentration and extraction of a number of pesticides such as aldrin, dieldrin, endrin, 4-dichlorodiphenyltrichloroethane (4-DDT) and heptachlor in natural water samples. Recently Wang et al [64] developed an electrochemical sensor using zirconia/ordered macroporous polyaniline and successfully employed for detecting the methyl parathion (organophosphate pesticides). An electrochemical sensor made from glassy carbon (GC) electrode, improved by a combination of multi-walled carbon nanotubes (MWCNT) with β-cyclodextrin (β-CD) incorporated in a polyaniline film, was also developed and tested to monitor the pyrimethanil fungicide in pome fruit [65]. A biosensor based on polyaniline (PANI) and multiwalled carbon nanotubes (MWCNT) for pesticides have been developed and demonstrated for determination of carbamate [66].

4.3 PANI-based adsorbents for the treatment of heavy metals contaminated wastewater

The different metals and their compounds released from the industry and other human activities into the environment and particularly into water resources strongly affect the natural properties of water. Although some of these metals are advantageous in biological processes, at higher concentrations, they become toxic to aquatic life and undesirable for humans. The most poisonous among the trace elements are the heavy metals and metalloids such as Hg, Pb, Cd, Ni, Cr, As, Sb, Se, etc. These metals are an important class of pollutants in water and wastewater systems. The heavy metals that may be bound to the cell membranes interfere with the transport phenomenon across the cell wall. The pollution of the environment by toxic metals is a common problem and in most cases, the source of pollution is originating from industrial activities. The metallic contaminants are of substantial importance as they are non-biodegradable and once discharged they cannot be destroyed and can only be either diluted or transformed. Therefore, their removal is essential from the wastewaters before discharge. As discussed earlier, the adsorption technique is the best one for the removal of pollutants from wastewater and has been very successfully used for the treatment of toxic metals. The applications of PANI-based composite as an adsorbent for heavy metals treatment are discussed below by taking some representative examples.

The sorption characteristics of PANI/Hexagonal mesoporous silica nanocomposites for the treatment of Ni(II) from water was investigated by Javadian et al. [67]. They reported that the removal of Ni(II) by this adsorbent is feasible, spontaneous and associated with the absorption of heat and accordingly removal efficiency increased with temperature. While analyzing the data with the three models, namely, Morris–Weber, pseudo-second order and pseudo second order, it was suggested that the adsorption was controlled by pseudo-second-order model. PANI/graphene oxide nanocomposites were prepared by oxidation polymerization in presence of dilute graphene oxide solution and aniline monomer at the temperature of − 20°C and used for the removal of Cr(VI) from aqueous solution with a very high q_m of 1149.4 mg/g [68]. Chen et al. [69] have prepared flake-like polyaniline/montmorillonite (PANI/MMT) nanocomposites by polymerization using poly(2-acrylamido-2-methylpropanesulfonic acid) and a polymer acid as a dopant of PANI. Flake thickness and surface roughness of PANI/MMT composites were reported to be decreased with the increase of montmorillonite/aniline feeding ratio. This nanocomposite adsorbent was successfully used to remove Cr(VI) from water. The influence of various factors pH, contact time, metal concentration and adsorbent dose were optimized. The kinetics behind the adsorption was followed well by the pseudo-second order model. A high adsorption capacity (167.5 mg/g) was reported. A novel polyaniline/attapulgite (PANI/ATP) composite with a considerably higher BET surface area was synthesized by employing the polymerization of aniline on the attapulgite template and utilized for the effective for adsorption of Hg(II) from water [70]. A very high maximum mercury adsorption capacity of 824 mg/g was reported to be achieved. It has also been observed that the solution pH has a major influence on the adsorption. Further, the adsorption Hg(II) adsorption by PANI/ATP was fitted well with the Langmuir isotherm and pseudo-second-order kinetic model, indicating that the Hg(II) adsorption process was predominantly controlled by a chemical process.

Mansour et al [71] used PANI coated sawdust adsorbent for adsorptive treatment of Cd (II) ions from water by the batch method and optimized the different physicochemical factors such as pH, adsorbate concentration, adsorbent dosage level and equilibrium contact time. On the basis of equilibrium and time-dependent studies, it was established that the adsorption of Cd(II) follows Freundlich isotherm and pseudo-second order kinetics. PANI coated sawdust as adsorbent was also used for the effective removal of Cu(II) with an average adsorption capacity of 58.2 mg/g [72]. The application of polyaniline/multiwalled carbon nanotubes (PANI/MWCNT) composites were investigated for the removal of Pb(II) from aqueous solutions and suggested to be efficient promising magnetic materials for the preconcentration and removal of heavy metals from huge volumes of aqueous solutions in environmental pollution cleanup [73]. The equilibrium, kinetic, thermodynamic and desorption studies of cadmium and lead by polyaniline grafted cross-linked chitosan beads from aqueous solution were studied and the maximum adsorption capacity for Cd(II) and Pb(II) ions at a temperature of 45 °C from Langmuir model was found to be 145 mg/g and 114 mg/g respectively and kinetic data were best represented by the pseudo-second order model for both ions [74]. The selected adsorbate-adsorbent systems for the removal of various contaminants from water have been summarized in Table 1.

Table 1   Selected works on the treatment contaminants from water.

Adsorbate (contaminant)	Adsorbent	Best fit isotherm model	Best fit Kinetic model	Reference
Methyl orange	PANI/Sawdust	Langmuir	Pseudo-second order	27
Sulfonated anionic dyes	PANI/Potash alum	Langmuir	Pseudo-second order	28
Acid violet 19	PANI/Fe$_2$O$_3$	Langmuir, Freundlich	Pseudo-second order	29
Methyl orange	PANI/Polyvinyl pyrrolidone	Langmuir	NR	30
Congo red	PANI/Bismuth tungstate	Redlich-Peterson, Langmuir	Pseudo-second order	31
Methylene blue	PANI/Sawdust	Langmuir	NR	32
Methylene blue	PANI/Nickel ferrite	Langmuir, Freundlich	Pseudo-second order	33
Congo red, Coomassie Brilliant blue, Remazol Brilliant blue R	PANI/Chitosan	Langmuir	Pseudo-second order	34
Methylene blue	PANI/Silica	Langmuir	Pseudo-second order	35
Acid red 18	PANI/Rice husk	Langmuir	Pseudo-second order	36

Rhodamine B, Congo red	PANI/α-MoO$_3$	Langmuir	Pseudo-second order	37
Orange 16	PANI/Chitosan/ZnO	Langmuir	Pseudo-second order	38
Methylene blue	PANI/Polyvinyl alcohol/Clinoptilolite	Langmuir	Intraparticle diffusion model	39
Methyl orange	PANI/Glass plate	Langmuir, Freundlich	Pseudo-second order	40
Tartrazine	PANI/Fe$_2$O$_3$	Freundlich	Intraparticle diffusion model	41
Methylene blue	PANI	Langmuir	Pseudo-first order	42
	PANI/Nitroprusside	Langmuir	Pseudo-second order	
Ni(II)	Polyaniline/Hexagonal Mesoporous Silica	Freundlich	Pseudo-second order	67
Cr(VI)	PANI, PANI/ Graphene oxide	Langmuir, Freundlich	Pseudo-second order	68
Cr(VI)	PANI/Montmorillonite	Langmuir, Freundlich,Temkin	Pseudo-second order	69
Hg(II)	PANI/Attapulgite	Langmuir	Pseudo-second order	70
Cd(II)	PANI/Sawdust	Freundlich	Pseudo-second order	71
Cd(II), Pb(II)	PANI/Chitosan	Langmuir	Pseudo-second order	74

5.   Conclusions

The contamination of ground water by dyes and heavy metals ions has become a universal environmental problem and the safe and effective disposal of industrial wastewater is thus a challenging task. Among the various treatment technologies adsorption technique has been found to be the most efficient and superior technique because of the advantages associated with it i.e. low cost, insensitivity to toxic pollutants, simplicity and flexibility of design. The present survey indicates the PANI-based composites are good substitutes of conventional adsorbents and are very successful for the removal of a variety of pollutants in general and dyes and heavy metals especially. It was generally observed that in most cases the adsorptive removal of toxic pollutants obeys the Langmuir isotherm and the kinetics behind the removal are best represented by the pseudo-second order model. Further, these PANI based composite materials have also been successfully used for monitoring the presence of pesticides in water.

Acknowledgements

The author is thankful to the chairman, Department of Applied Chemistry, Faculty of Engineering and Technology, Aligarh Muslim University, Aligarh, India, for providing the necessary facilities.

References

[1]   E.A. Laws, Aquatic Pollution: An Introductory Text, 3rd ed.; John Wiley & Sons: New York, 2000.

[2]   H.M. Pignon, C.F. Brasquet, P.L. Cloirec, P. L. (2003). Adsorption of dyes onto activated carbon cloths: Approach of adsorption mechanisms and coupling of ACC with ultrafiltration to treat coloured wastewaters, Sep. Purif. Technol., 31 (2003) 3–11. https://doi.org/10.1016/S1383-5866(02)00147-8

[3]   A. Nasar A, S. Shakoor, Remediation of dyes from industrial wastewater using low-cost adsorbents. In "Applications of Adsorption and Ion Exchange Chromatography in Waste Water Treatment" Inamuddin, Al-Ahmed A (eds), Materials Research Foundations, Vol. 15 (2017) 1-33. https://doi.org/10.21741/9781945291333-1

[4]   S. Shakoor, A. Nasar, Removal of methylene blue dye from artificially contaminated water using citrus limetta peel waste as a very low cost adsorbent, J. Taiwan Inst. Chem. Eng. 66 (2016), 154–163. https://doi.org/10.1016/j.jtice.2016.06.009

[5]   S. Shakoor, A. Nasar, Adsorptive treatment of hazardous methylene blue dye from artificially contaminated water using cucumis sativus peel waste as a low-cost adsorbent, Groundwater Sustain. Develop., 5 (2017) 152–159. https://doi.org/10.1016/j.gsd.2017.06.005

[6]   V. Vimonses, B. Jin, C.W.K. Chow, Insight into removal kinetic and mechanisms of anionic dye by calcined clay materials and lime, J. Hazard. Mater., 177 (2010) 420–427. https://doi.org/10.1016/j.jhazmat.2009.12.049

[7]   T. Robinson, G. Mcmullan, R. Marchant, P. Nigam, Remediation of dyes in textiles effluent: A critical review on current treatment technologies with a proposed alternative, Bioresour. Technol. 77 (2001) 247–255. https://doi.org/10.1016/S0960-8524(00)00080-8

[8]   P.B. Fai, A. Grant, A rapid resazurin bioassay for assessing the toxicity of fungicides, Chemosphere., 74 (2009) 1165-70. https://doi.org/10.1016/j.chemosphere.2008.11.078

[9]   T. Perez-Ruiz, C. Martinez-Lozano, A. Sanz, E. Bravo, Determination of organophosphorus pesticides in water, vegetables and grain by automated SPE and MEKC, Chromatographia, 61 (2005) 493–498. https://doi.org/10.1365/s10337-005-0533-8

[10]   R. Rezg, B. Mornagui, S. El-Fazaa, N. Gharbi, Organophosphorus pesticides as food chain contaminants and type 2 diabetes: A review, Trends Food Sci. Technol. 21 (2010) 345–357. https://doi.org/10.1016/j.tifs.2010.04.006

[11]   J. Fenik, M. Tankiewicz, M. Biziuk, Properties and determination of pesticides in fruits and vegetables, Trends Anal. Chem. 30 (2011) 814–826. https://doi.org/10.1016/j.trac.2011.02.008

[12]   A. Agrawal, R.S. Pandey, B. Sharma, water pollution with special reference to pesticide contamination in India, J. Water Resour. Protect. 2 (2010) 432-448. https://doi.org/10.4236/jwarp.2010.25050

[13]   T. Ahmad, M. Rafatullah, A. Ghazali, O. Sulaiman, R. Hashim, A. Ahmad, Removal of pesticides from water and wastewater by different adsorbents: a review, Journal of Environmental Science and Health, Part C, 28 (2010) 231-271. https://doi.org/10.1080/10590501.2010.525782

[14]   W. Mathys, Pesticide pollution of groundwater and drinking water by the processes of artificial groundwater enrichment or coastal filtration: underrated sources of contamination, Zentralb. Hyg. Umweltmed., 196 (1994) 338-359.

[15]   G.H. Willis, L. L. McDowell, Pesticides in agricultural runoff and their effects on downstream water quality, Environ. Toxicol. Chem. 1 (1982) 267-219. https://doi.org/10.1002/etc.5620010402

[16]   A.B.P. Marin, V.M. Zapata, J.F. Orturao, M. Aguilar, M.; J. Saez, M. Lloren, Removal of cadmium from aqueous solutions by adsorption on to orange waste, J. Hazard. Mater. 139 (2007) 122-131. https://doi.org/10.1016/j.jhazmat.2006.06.008

[17]   F.S. Zhang, J.O. Nriagu, H. Itoh, Mercury removal from water using activated carbons derived from organic sewage sludge, Water. Res. 39 (2005) 389-395. https://doi.org/10.1016/j.watres.2004.09.027

[18]   F.S. Zhang, J.O. Nriagu, H. Itoh, Mercury removal from water using activated carbons derived from organic sewage sludge, Water. Res. 39 (2005) 389-395. https://doi.org/10.1016/j.watres.2004.09.027

[19]   G. Bayramoglu, M.Y. Arica, Kinetics of mercury ions removal from synthetic aqueous solutions using by novel magnetic p(GMA-MMA-EGDMA) beads, J. Hazard. Mater., 144(2007) 449-457. https://doi.org/10.1016/j.jhazmat.2006.10.058

[20]   Sigel, H.; Sigel, A. Concepts on Metal Ion Toxicity; Marcel Dekker, Inc.: New York, 1986.

[21]   I. Langmuir, The constitution and fundamental properties of solids and liquids. Part I. Solids., J. Am. Chem. Soc. 38 (1916) 2221–2295. https://doi.org/10.1021/ja02268a002

[22]   H.M.F. Freundlich, Über die adsorption in lösungen, Z. Phys. Chem., 57 (1906) 57, 385–470.

[23]   M.I. Temkin, V. Pyzhev, Kinetics of ammonia synthesis on promoted iron catalyst. Acta Phys. Chim. USSR., 12 (1940) 217-222.

[24]   T.A. Khan, S. Dahiya and I. Ali, Use of kaolinite as adsorbent: Equilibrium, dynamics and thermodynamic studies on the adsorption of Rhodamine B from aqueous solution. Appl. Clay Sci., 69 (2012) 69, 58–66.

[25]   F. Kanwal, R. Rehman, T. Mahmud, J. Anwar, R. Ilyas, Isothermal and thermodynamical modeling of chromium (III) adsorption by composites of polyaniline with rice husk and saw dust. J. Chil. Chem. Soc., 57 (2012) 1058-1063. https://doi.org/10.4067/S0717-97072012000100022

[26]   J. Li, Y. Huang, and D. Shao, Conjugated polymer-based composites for water purification, in Book entitled "Fundamentals of Conjugated Polymer Blends,

Copolymers and Composites: Synthesis, Properties and Applications" Ed. P. Saini, Scrivener Publishing LLC, 2015, pp 581-618.

[27]   R. Ansari, Z. Mosayebzadeh, Application of polyaniline as an efficient and novel adsorbent for azo dyes removal from textile wastewaters, Chemical Pap. 65 (2011) 1 -8.

[28]   B.N. Patra, D. Majhi, Removal of anionic dyes from water by potash alum doped polyaniline: investigation of kinetics and thermodynamic parameters of adsorption, J. Phys. Chem. B, 119 (2015) 8154-8164. https://doi.org/10.1021/acs.jpcb.5b00535

[29]   M.R. Patil, V.S. Shrivastava, Adsorption removal of carcinogenic acid violet 19 dye from aqueous solution by polyaniline-Fe2O3 magnetic nano-composite, J. Mater. Environ. Sci., 6 (2015) 11-21.

[30]   A. R. Prasad, A. Joseph, Synthesis, characterization and investigation of methyl orange dye removal from aqueous solutions using waterborne poly vinyl pyrrolidone (PVP) stabilized poly aniline (PANI) core–shell nanoparticles, RSC Adv., 7 (2017) 20960-20968. https://doi.org/10.1039/C7RA01790A

[31]   M. Laabd, H. A. Ahsaine, A. E. Jaouhari, B. Bakiz, M. Bazzaoui, M. Ezahri, A. Albourine, A. Benlhachemi, Congo red removal by PANI/Bi2WO6 nanocomposites: Kinetic, equilibrium and thermodynamic studies, J. Environ. Chem. Eng., 4 (2016) 3096-3105. https://doi.org/10.1016/j.jece.2016.06.024

[32]   M.B. Keivani, K. Zare, H. Aghaie, R. Ansari, Removal of methylene blue dye by application of polyaniline nano composite from aqueous solutions, J. Phys. Theor. Chem. IAU, 6 (2009) 50-56.

[33]   M.R. Patil, V.S. Shrivastava, Adsorptive removal of methylene blue from aqueous solution by polyaniline-nickel ferrite nanocomposite: a kinetic approach, Desal. Water Treat., 57 (2016) 5879-5887. https://doi.org/10.1080/19443994.2015.1004594

[34]   V. Janaki, B.T. Oh, K. Shanthi, K.J. Lee, A.K. Ramasamy, S. Kamala-Kannan, Polyaniline/chitosan composite: an eco-friendly polymer for enhanced removal of dyes from aqueous solution, Synth. Met., 162 (2012) 974-980. https://doi.org/10.1016/j.synthmet.2012.04.015

[35]   M.M. Ayad, A.A. El-Nasr, J. Stejskal, Kinetics and isotherm studies of methylene blue adsorption onto polyaniline nanotubes base/silica composite, J. Ind. Eng. Chem. 18 (2012) 1964-1969. https://doi.org/10.1016/j.jiec.2012.05.012

[36] M. Shabandokht, E. Binaeian, H.A. Tayeb, Adsorption of food dye Acid red 18 onto polyaniline-modified rice husk composite: isotherm and kinetic analysis, Desal. Water Treat., 57 (2016) 27638-27650. https://doi.org/10.1080/19443994.2016.1172982

[37] S. Dhanavel, E.A.K. Nivethaa, K. Dhanapal, V.K. Gupta, V. Narayan, A. Stephen, α-MoO3/polyaniline composite for effective scavenging of Rhodamine B, Congo red and textile dye effluent, RSC Adv., 34 (2016) 28871-28886. https://doi.org/10.1039/C6RA02576E

[38] P. Kannusamy, T. Sivalingam, Synthesis of porous chitosan–polyaniline/ZnO hybrid composite and application for removal of reactive orange 16 dye, Colloid. Surf. B: Biointerfac. 108 (2013) 229-238. https://doi.org/10.1016/j.colsurfb.2013.03.015

[39] A. Rashidzadeh & A. Olad, Novel polyaniline/poly (vinyl alcohol)/clinoptilolite nanocomposite: dye removal, kinetic, and isotherm studies, Desal. Water Treat., 51 (2013) 7057-7066. https://doi.org/10.1080/19443994.2013.766904

[40] K. Haitham, S. Razak, M.A. Nawi, Kinetics and isotherm studies of methyl orange adsorption by a highly recyclable immobilized polyaniline on a glass plate, Arab. J. Chem. 2014 (In press) https://doi.org/10.1016/j.arabjc.2014.10.010 https://doi.org/10.1016/j.arabjc.2014.10.010

[41] S. A. Jebreil, Removal of tartrazine dye form aqueous solutions by adsorption on the surface of polyaniline/iron oxide composite, Int. J. Chem. Mol. Nucl. Mater. Metall. Eng. 8 (2014) 1433-1438.

[42] F. A. Rafiqi, K. Majid, Sequestration of methylene blue (MB) dyes from aqueous solution using polyaniline and polyaniline–nitroprusside composite, J. Mater. Sci. 52 (2017) 6506–6524. https://doi.org/10.1007/s10853-017-0886-z

[43] M. Mahalakshmi, M. Palanichamy, A. Banumathi, V. Murugesan, 'Photocatalytic degradation of carbofuran using semiconductor oxides', J. Hazard. Mater. 143 (2007) 240-245. https://doi.org/10.1016/j.jhazmat.2006.09.008

[44] T. Aungpradit, P. Sutthivaiyakit, D. Martens, S. Sutthivaiyakit, A.A.F. Kettrup, Photocatalytic degradation of triazophos in aqueous titanium dioxide suspension: Identification of intermediates and degradation pathways. J. Hazard. Mater., 146 (2007) 204–213. https://doi.org/10.1016/j.jhazmat.2006.12.007

[45] P. Saritha, C. Aparna, V. Himabindu, Y. Anjaneyulu, Comparison of various advanced oxidation processes for the degradation of 4-chloro-2 nitrophenol, J.

Hazard.Mater., 149 (2007) 609–614.
https://doi.org/10.1016/j.jhazmat.2007.06.111

[46]   A.L. Ahmad, L.S. Tan, S.R.A. Shukor, Dimethoate and atrazine retention from
        aqueous Solution by nanofiltration membranes, J. Hazard. Mater., 151 (2008) 71–
        77. https://doi.org/10.1016/j.jhazmat.2007.05.047

[47]   M.H.M. Rajashekara, H.K. Manonmani, Aerobic degradation of technical
        hexachlorocyclohexane by a defined microbial consortium, J. Hazard. Mater., 149
        (2007) 18–25. https://doi.org/10.1016/j.jhazmat.2007.03.053

[48]   C. Masselon, G. Krier, J.F. Muller, S. Nelieu, J. Einhorn, Laser desorption Fourier
        transformation cyclotron resonance mass spectrometry of selected pesticides
        extracted on C18 silica solid-phase extraction membranes. Analyst. 121 (1996)
        1429–1433. https://doi.org/10.1039/AN9962101429

[49]   A.J.M. Lagadec, D.J. Miller, A.V. Lilke, S.B. Hawthorne, Pilot-scale subcritical
        water remediation of polycyclic aromatic hydrocarbons- and pesticide-
        contaminated soil, Environ. Sci. Technol. 34 (2000) 1542–1548.
        https://doi.org/10.1021/es990722u

[50]   Qamruzzaman, A. Nasar, Degradation of tricyclazole by colloidal manganese
        dioxide in the absence and presence of surfactants. J. Ind. Eng. Chem. 20 (2014)
        897–902. https://doi.org/10.1016/j.jiec.2013.06.020

[51]   Qamruzzaman, A. Nasar, Treatment of acetamiprid insecticide from artificially
        contaminated water by colloidal manganese dioxide in the absence and presence of
        surfactants, RSC Adv. 4 (2014) 62844–62850.

[52]   Qamruzzaman, A. Nasar, Kinetics of metribuzin degradation by colloidal
        manganese dioxide in absence and presence of surfactants. Chemical Pap. 68
        (2014) 65–73.

[53]   Qamruzzaman, A. Nasar, Degradation of acephate by colloidal manganese dioxide
        in the absence and presence of surfactants. Desal. Water Treat., 55 (2015) 2155–
        2164. https://doi.org/10.1080/19443994.2014.937752

[54]   Qamruzzaman, A. Nasar, Degradation of methomyl by colloidal manganese
        dioxide in acidic medium, Chem. Sci. Rev. Lett., 1 (2012) 113-119.

[55]   M.M.B. Martın, J.A.S. Perez, J.L.G. Sanchez, L.M. Oca, J.L.C. Lopez, I. Oller,
        S.M. Rodrıguez, Degradation of alachlor and pyrimethanil by combined photo-
        fenton and biological oxidation. J. Hazard. Mater., 155 (2008) 342-349.
        https://doi.org/10.1016/j.jhazmat.2007.11.069

[56] V.F. Dominquis, G. Priolo, A.C. Alves, M.F. Cabral, C. Deleure-Matos, Adsorption behavior of α-cypermethrin on cork and activated carbon, J Environ. Sci. Health B, 42 (2007) 649-654. https://doi.org/10.1080/03601230701465635

[57] M. Akhtar M, S.M. Hasany, M.I. Bhanger, S. Iqbal, Low cost sorbents for the removal of methyl parathion pesticide from aqueous solutions, Chemosphere, 66 (2007) 1829-1838. https://doi.org/10.1016/j.chemosphere.2006.09.006

[58] G.Z. Memon, M.I. Bhanger, M. Akhtar, The removal efficiency of chestnut shells for selected pesticides from aqueous solutions. J Colloid Interface Sci., 315 (2007) 33-40. https://doi.org/10.1016/j.jcis.2007.06.037

[59] G.Z. Memon, M.I. Bhanger, M. Akhtar, F.N. Talpur, J.R. Memon, Adsorption of methyl parathion pesticide from water using watermelon peels as a low cost adsorbent, Chem. Eng. J., 138 (2008) 616-621. https://doi.org/10.1016/j.cej.2007.09.027

[60] M.S. Rodriguez-Cruz, M.S. Andrades, A.M. Parada, M.J. Sanchez-Martin, Effect of different wood pretreatments on the sorption-desorption of linuron and metalaxyl by woods, J. Agric. Food. Chem., 56 (2008) 7339-7346. https://doi.org/10.1021/jf800980w

[61] J. Ludvíka, P. Zuman, Adsorption of 1, 2, 4-triazine pesticides metamitron and metribuzin on lignin, Microchem. J., 64 (2000) 15-20. https://doi.org/10.1016/S0026-265X(99)00015-6

[62] X. Yang, K. Qiao, Y. Ye, M. Yang, J. Li, H. Gao, S. Zhang, W. Zhou, R. Lu, Facile synthesis of multifunctional attapulgite/Fe3O4/polyaniline nanocomposites for magnetic dispersive solid phase extraction of benzoylurea insecticides in environmental water samples, Anal. Chim. Acta., 934 (2016) 114-121. https://doi.org/10.1016/j.aca.2016.06.027

[63] A. Mehdinia, Preconcentration and determination of organochlorine pesticides in seawater samples using polyaniline/polypyrrole-cellulose nanocomposite-based solid phase extraction and gas chromatography-electron capture detection, J. Braz. Chem. Soc., 25 (2014) 2048-2053. https://doi.org/10.5935/0103-5053.20140190

[64] Y. Wang, J. Jin, C. Yuan, F. Zhang, L. Ma, D. Qin, D. Shan, X. Lu, A novel electrochemical sensor based on zirconia/ordered macroporous polyaniline for ultrasensitive detection of pesticides, Analyst, 140 (2015) 140, 560-566.

[65] J.M.P.J. Garrido, V. Rahemi, F. Borges, C.M.A. Brett, E.M.P.J. Garrido, Carbon nanotube β-cyclodextrin modified electrode as enhanced sensing platform for the

determination of fungicide pyrimethanil, Food Control, Volume 60, February 2016, Pages 7-11. https://doi.org/10.1016/j.foodcont.2015.07.001

[66] I. Cesarino, F.C. Moraes, S.A. Machado, a biosensor based on polyaniline-carbon nanotube core-shell for electrochemical detection of pesticides, Electroanal. 23 (2011) 2586-2593. https://doi.org/10.1002/elan.201100161

[67] H. Javadian, P. Vahedian, M. Toosib, Adsorption characteristics of Ni(II) from aqueous solution and industrial wastewater onto Polyaniline/HMS nanocomposite powder, Appl. Surf. Sci., 284 (2013) 13-22. https://doi.org/10.1016/j.apsusc.2013.06.111

[68] S. Zhang, M. Zeng, W. Xu, J. Li, J. Li, J. Xu, X. Wang, Polyaniline nanorods dotted on graphene oxide nanosheets as a novel super adsorbent for Cr(VI)., Dalton Trans., 42 (2013) 7854-7858. https://doi.org/10.1039/c3dt50149c

[69] J. Chen, X. Q. Hong, Y. T. Zhao, Y. Y. Xia, D. K. Li, Q. F. Zhang, Preparation of flake-like polyaniline/montmorillonite nanocomposites and their application for removal of Cr(VI) ions in aqueous solution, J. Mater. Sci., 48 (2013) 7708-7717. https://doi.org/10.1007/s10853-013-7591-3

[70] H. Cui, Y. Qian, Q. Li, Q. Zhang, J. P. Zhai, Adsorption of aqueous Hg(II) by a polyaniline/attapulgite composite, Chem. Eng. J., 211 (2012) 216-223. https://doi.org/10.1016/j.cej.2012.09.057

[71] M. S. Mansour, M. E. Ossman, H. A. Farag, Removal of Cd (II) ion from waste water by adsorption onto polyaniline coated on sawdust, Desalination, 272 (2011) 301-305. https://doi.org/10.1016/j.desal.2011.01.037

[72] D. L. Liu, D. Z. Sun, Modeling adsorption of Cu(II) using polyaniline-coated sawdust in a fixed-bed column. Environ. Eng. Sci., 29 (2012) 461-465. https://doi.org/10.1089/ees.2010.0435

[73] D. D. Shao, C. L. Chen, X. K. Wang, Application of polyaniline and multiwalled carbon nanotube magnetic composites for removal of Pb(II), Chem. Eng. J., 185 (2012) 144-150. https://doi.org/10.1016/j.cej.2012.01.063

[74] E. Igberase, P. Osifo, Equilibrium, kinetic, thermodynamic and desorption studies of cadmium and lead by polyaniline grafted cross-linked chitosan beads from aqueous solution, J. Indust. Eng. Chem., 26 (2015) 340-347. https://doi.org/10.1016/j.jiec.2014.12.007

Chapter 3

Smart Polymeric Coatings to Enhance the Antibacterial, Anti-fogging and Self-Healing Nature of a Coated Surface

Santanu Sarkar[1], Chiranjib Bhattacharjee[2]*, Supriya Sarkar[1]

[1] Environment Research Group, R&D, Tata Steel Ltd., Jamshedpur-831007, India

[2] Chemical Engineering Department, Jadavpur University, Kolkata-700032, India

* cbhattacharyya@chemical.jdvu.ac.in, c.bhatta@gmail.com

Abstract

All surface modifications are mainly done to improve different properties e.g. antibacterial, anti-fogging, self-healing etc. of a surface. By definition '*smart coating*' is a type of coating which has such special properties. The current chapter deals with three special characteristics of such smart coatings. Antibacterial coatings make the surface immune to living microorganisms and thus the growth of bacteria on such coated surface becomes restricted. In the case of self-healing coatings, the surface is protected from a corrosive environment and moreover, if the coating is damaged due to some reason self-repair mechanism prevents the exposed surface from any type of attack. Finally, anti-fogging coatings enhance the hydrophilic nature of the coated surface as well as reduce any light penetration problem through the surface by preventing water droplet formation on the coated surface in moist weather. All types of coatings have a large amount of applications in different fields of science as well as in our daily lives. This chapter has also highlighted several aspects and applications of different types of smart coatings.

Keywords

Smart Coating, Antibacterial Coating, Anti-fogging Coating, Self-Healing Coating, Surface Modification

Contents

1. Introduction

The surface of any material is coated with other material to fulfill a certain purpose. For example, different structures are coated with paints to limit the corrosion as well as increase the lifetime of those structures. Smart coating belongs to a similar category but by definition, this type of coating is applied on a specific surface to incorporate some unique properties on that particular surface. These properties are mainly antibacterial, anti-fogging and self-healing properties. All of these properties may exist in the same material or individually. Several research works on smart coatings have been carried out to fulfill various types of requirements.

1.1 Antibacterial coatings

In the field of medical science, the most significant thing is to maintain antibacterial condition during the whole treatment process. A great percentage of total health-care cost is being utilized to maintain hygienic conditions [1-4]. Significant bacterial infections are observed during the implementation of medical devices e.g. in Europe alone it has been reported that approximately 800,000 orthopedic implantations take place annually among that 1.5% suffers infections after the treatment [1]. Thus, these types of infections cause multiple operations and health care costs as well as morbidity rate increase [1, 4-6]. Moreover, some major infections due occur during regular intervention and implantation of a foreign body. Center for Disease Control reported that during the treatment of breast cancer an overall of 2% surgical site infection (SSI) occurre [7] but the individual rate varied between 1% to 28% and the rate of SSI was more pronounced in the case of mastectomy with immediate breast implantation [7]. Therefore, synthetic implantation enhances the rate of infection. Another major infection in blood stream occurs during the use of central venous catheter [8-10]. To recover from such type of infection it is mandatory to stay in the hospital for a longer time and thus health care cost are high [11, 12]. Therefore, this serious problem caused by bacterial activity can be counter attacked not only by improving the knowledge of staff but with the implementation of safety and efficient medical devices to reduce the contamination efficiently [12, 14-16]. Detailed ways of improvement with help of antibacterial smart coating have been demonstrated in the present chapter.

1.2 Anti-fogging coatings

Smart coatings that work as anti-fogging coatings are a typically thin layer that prevents the formation of water droplet on that particular coated surface. The coating is hydrophilic in nature, and therefore the film causes moisture to spread into an even layer without the formation of droplets, which make an obstacle for clear vision. By maintaining clear visibility anti-fogging coating reduces the chances of unusual casualties. Such type of coatings is applied on glasses, goggles, camera lenses, binoculars etc. and more importantly, they are introduced into automobile industries. No such literature is available on this current topic but an elaborate description on anti-fogging coating has been provided in the later part of the present work.

1.3 Self-healing coatings

The lifetime of any essential item is important. The surface of any metal structure made of steel and magnesium/aluminum alloys are affected by the environment. To prolong the lifetime of those materials, surfaces are coated with some foreign materials which can be done with the help of immersion and electrochemical deposition or using modern equipment like PVD, CVD and laser plating etc. In recent scenarios, the coating not only protects metals from the harmful environment but it also recovers damages caused by the working environment to regain their original properties. This type of coating is called self-healing smart coating and self-healing is the most important property of the coating. It has been reported, during 2000 to 2010, that the use of self-healing materials has increased up to ten times [16]. The main self-healing anticorrosion coating of chromium (VI) compounds have been first introduced but due to restriction in use of Cr(VI) it has been replaced by some other coating material. Nowadays, macromolecular compounds, ceramics and metals composites are frequently used as self-healing coatings along with cerium, sodium tetraoxomolybdate, colloidal silica and fluoro-organic compounds. The properties of those coatings are enhanced by changing temperature, radiation, pH, pressure as well as by applying mechanical actions. In the current context, the development and application of anti-corrosion and anti-healing coating in a different environment as well as on different metals have been illustrated.

2. Antibacterial coating: aspects, methods and applications

2.1 Sources of infections

It is important to identify the causes and source of infections to nullify these unwanted means. According to Rosenthal et al. [17] and Bearman et al. [18] there are several drawbacks existing in health care units, which causes infections. Those are represented as contaminated surfaces like implant/device surface, hands of the surgical staff during implantation/application, patient's own skin or mucus membrane and contaminated disinfectants as well as contact with other infected patients in the hospital, or family members after the intervention. However, it is not possible to avoid all types of infections but the patient should be shielded from such contaminations as much as possible. Medical devices are generally delivered to the hospital under proper sterilized condition but infections spread due to miss handling [19].

During any treatment, proteins from the blood or tissue are adsorbed onto medical devices and the rate of adsorption depends on the surface hydrophobicity, roughness, porosity, chemical composition, as well as composition and concentration of the protein solution, salt concentrations, pH, etc. However, it is impossible to identify the type of protein, which has a better adsorption on the surface of devices [20-22]. On this adsorbed protein, a layer of free-swimming bacteria are attached in the planktonic state [23, 24] and therefore they increase in number by using the protein as source to form a colony of bacteria. More importantly, they form a biofilm with extracellular polymeric substances, which work as a protective layer, and they spread themselves throughout the system [24, 25]. In this way, the medical instruments are being contaminated. The way of contamination is represented by Fig. 1.

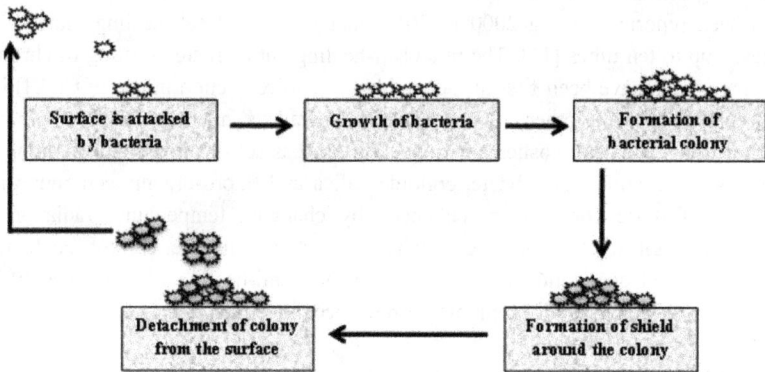

Fig. 1 The schematic representation of the steps involves in bacterial contamination of medical instruments.

It is very tough to break that biofilm by any means and it is next to impossible to kill these bacteria. Thus the use of biofilm disturbing agents along with antibiotics can be applied but it cannot be used frequently due to the harmful effects of the chemicals involved [24]. It is important to develop new strategies to eradicate both contamination and biofilm formation. Smart coating has been adopted to fulfill this purpose and the most significant thing during the development of such materials is that those materials should be compatible with the surrounding tissues.

2.2 Antibacterial coatings: types and applications

2.2.1 Surface modification

The growth of biofilm formation on the surfaces of medical implants is very common and there are so many reasons for the formation of such a type of film, which has already been discussed above. Therefore, it is important to develop a surface, which has antibacterial properties. This can be done with surface modification having anti-microbial properties. The smooth surface is less susceptible to the biofilm growth by free-swimming pathogens as smooth surface poses lesser adhesive force and better hydrophobic property than a rough surface and at the same time a rough surface provides a larger surface area for bacterial growth [23, 26-28]. Surprisingly however, microorganisms have a charged outer surface along with hydrophobic active sites and thus they can grow on hydrophobic surfaces. Hence the growth of bacteria always depends on the formation of a protein layer on the surface of medical instruments [23, 24].

There are several strategies available for antibacterial smart coatings like surface modification with chemicals having protein & bacteria-repelling property, quaternary ammonium, antibiotics and noble metals, especially silver.

2.2.1.1 Protein and bacteria-repelling coatings

Polymer coatings on medical devices are the most widespread techniques used in health care systems. Several types of hydrophilic polymers have been used i.e. linear, branched, star-shaped etc. and those might be homopolymers or block copolymers [29-31]. The polymers are attached to the surface by noncovalent interactions e.g. the interaction between the sulfur atoms with the gold surface and this interaction can be further improved in presence of thiol groups, which work as a linker. The acrylate or methacrylate function helped the polymers to attach to the gold surface, which was done by grafting technique [32,33]. A block copolymer, poly(ethyleneglycol)-bl-poly(propylene sulphide)-bl- poly(ethylene glycol) contain thiol group, forms a stable and tight coating on the gold surfaces [34]. This technique of sulfur to gold strategy has been proven as an advantage because of the formation of a monolayer on the metal surface and this technique is easy to perform. However, it has some drawbacks like instability and uncontrolled formation of monolayers [32, 33]. Not only gold but also titanium can be coated with a polymer brush using peptide linked PEG as the peptides have a higher affinity to titanium surfaces. Such type of coated surfaces showed better repelling property [35].

A certain dense thickness of polymer layer is maintained on the metal surface and this thickness depends on the chain length of the applied polymer, larger chain length results in a thicker polymeric layer. Denser and thicker layers showed a better anti-adhesive property. Therefore a longer chain polymer is favorable for the generation of antibacterial coating on metal surfaces [36-38]. Due to hydrophilic nature of the polymer, water molecules are attached with brush layer by hydrogen bonding. Then steric hindrance presents on the layer makes the surface repelling in nature for proteins and bacteria. This phenomenon is represented schematically in Fig. 2.

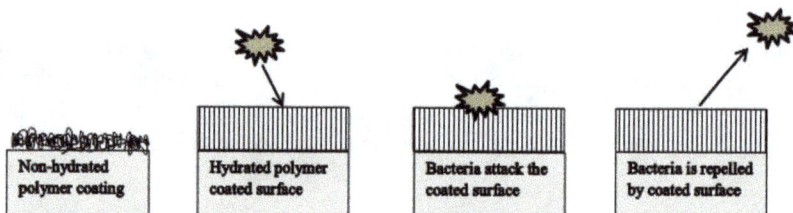

Fig. 2 The mechanism of the bacteria-repelling tendency of the polymer-coated surface.

2.2.1.2   Surface modification with quaternary ammonium compounds

The antibacterial properties of any surface can be improved by altering the polarity of the surface and a positively charged surface restricts bacterial activity. Quaternary ammonium, an antimicrobial compound, provides positive polarity to any surface, thus in several cases, it has been used for coating purpose to prepare anti-microbial surfaces. [39]. Quaternary ammoniums are water soluble and have several applications for disinfection purpose i.e. water and wastewater treatment and also in industrial application like a preserving agents in cosmetic products. Those also have antimicrobial activity during contact with a solid surface [40-43]. These antibacterial agents also help to remove vegetative bacteria, yeast, viruses, algae, fungi etc. but they cannot destroy bacterial spores, mycobacterium and hydrophilic viruses [44, 45]. Quaternary ammoniums have characteristics similar to those of cationic surfactants, which work as antibacterial surfactants.

The wide applicability of quaternary ammonium as an antimicrobial agent is because of its huge production. Moreover, it has higher stability in the human body and after use, it comes out from human body during excretion without any type alteration in both itself and living beings. The only disadvantage of this type of ammonium is that it is neither cytocompatible nor biocompatible [46]. Considering the advantageous areas present by this type of coating has a vast application where direct contact between the surface and skin occurs e.g. textiles, floors, and ceilings in a hospital. Bacteria die immediately after making contact with a quaternary ammonium coated surface. Although, long time exposure cause evolution of the cell membrane of the bacteria. Thus, the efficiency of such antimicrobial agent is reduced to some extent [46]. That is the major limitation of use of quaternary ammoniums.

2.2.1.3   Smart coatings with antibiotics

The biofilm on medical devices generated by microorganisms is very tough to break by antibacterial actions therefore, it is necessary to destroy the biofilm before its formation. In orthopedic surgery and orthodontic implants, local antibiotics are used frequently [48, 49]. The infection generally spread out after few days of surgery and to avoid this situation some antibiotic like gentamycin is mixed with the cement, which is used for fixing bones. This way the local growth of bacteria is inhibited by the cement itself because it releases antibiotic slowly after hardening.

Sometime biodegradable materials are used for coatings of implantations and antibiotics are loaded inside as biodegradable component. Thus, during degradation of the coating, the antibiotic is released and the surface becomes free from contamination of microbial attack. It is necessary that the rate of degradation and rate of release of antibiotic both are the same and such type of application is very common for orthopedic treatments [50, 51]. Some major coatings containing antibiotics are poly (D,L-lactide) (PDLLA) containing gentamycin, a mixture of rifampicin and fusidic acid in PLLA coatings [52] and chlorohexidine antibiotic in a variety of surface coatings like PVP, polyurethane, polyphosphoester (Politerefate), calcium phosphate. However, the calcium phosphate and PVP coatings release insufficient antibiotic thus the coated surface failed to prevent infection [53].

A regular use of hydroxyapatite as a coating material in orthopedic operation is observed because this type of material boosts osseointegration which results in earlier bone formation. Some antibiotics have calcium-chelating property and thus antibiotics can sustain longer period on bone tissues. The coating of such types of antibiotics was done by co-precipitation implant surfaces with help of hydroxyapatite thus; the surfaces could release antibiotics as per requirements. [54]. Moreover, the direct antibacterial coating on the implant surfaces using a proper solvent of that antibiotic can enhance the performance of it but sometimes this type of coating hampers local tissues due to its higher concentration. During the coating process, the antibiotics are attached to the surface by evaporating its solvents [55].

Major infections can take place during the use of a catheter, thus it is necessary to develop such type of catheter so that it can protect microbial activity. To fulfill that requirement, both internal and external surfaces of the catheter have been coated with a combination of antibiotic and antiseptic agents like minocycline and rifampin or chlorohexidine and silver-sulfadiazine [56, 57]. According to experimental observation, the antibacterial coating is more effective and economical than an antiseptic coating on catheters [11, 58]. Though this type of coated catheter is not very effective for long-term use because it loses its antimicrobial characteristic and thus it becomes susceptible to microbial growth [59]. Therefore, continuous release of antibiotics is required to solve this problem but at the same time the growth of antibiotic resistant bacteria can be observed and those bacteria cannot be removed further using general antibiotics. Finally, it may be concluded that direct use of antibiotics on implant surfaces is not a permanent solution to restrict local infections caused by microbial actions [60, 61].

2.2.2 Smart coatings: silver and silver nano-particle

2.2.2.1 Antimicrobial effect of silver and silver salts

Silver is a well-known antibacterial agent for a long time and it has wide applications in water and food preservation [62, 63]. In addition, silver and silver salts are used in the treatment of wounds from the ancient age to current time in medical treatments [64, 65]. The silver sutures and silver nitrate were widely used for the treatment before the discovery and introduction of antibiotics in medical treatment [62]. The silver ion (Ag^+) is responsible for the antibacterial activity [66]. Though, after the introduction of antibiotic the use of silver or silver salt was reduced to a greater extent, $AgNO_3$ was again used as silver sulfadiazine cream to treat burn wounds. Surprisingly it reduced the rate of infection but the healing became slow and therefore it is required to develop other alternative materials, which can be used as a replacement for antibiotics [67].

Nowadays silver nanoparticles are used as a replacement of silver and such type of nanoparticles have good antibacterial property. The working principle of Ag nanoparticles is still under research and thus improper. Some in vitro studies have already reported that the use of highly concentrated silver nanoparticles causes cell death.

Though recently silver nanoparticle coatings have been used for several applications like device coatings, wound dressings solution, deodorants and cosmetics [67, 68]. Silver nanoparticles have been used for surface coating of medical devices as a replacement of metallic silver as normal silver has poor antibacterial activity. Moreover, Ag salt cannot be used for coating purpose, as the stability of those salts is very poor [64,67,69].

2.2.2.2 Surface coatings using silver or silver nanoparticles

Several studies had been reported regarding the effective use of silver sulfadiazine cream or another medicine containing silver and silver salt for the treatment of burns and non-healing wounds [70, 71] and such type of dressing showed a better result than other types of wound dressing. The healing rate of a wound depended on the type of wound and type of dressing solution used during the treatment [71]. In the case of a burn wound, nano-silver was more effective than other silver formulations (silver sulfadiazine and silver nitrate) [72] and research showed that for the efficient treatment of deep burn wound nanocrystalline silver impregnated dressings was more suitable [73]. Therefore, nano silver is more effective than metallic silver.

In recent years, coatings containing silver nanoparticles have been used to stop infection like ventriculomeningitis. It has been reported that the colonization of the drains was reduced by a factor of four and the infection of the central spinal fluid by a factor of two [74]. In the case of catheter infection, several studies showed that nanosilver coated catheters could be implanted in an animal body successfully [75, 76]. The rate of infection due to the use of urinary catheters could be controlled significantly using silver coated catheters [77,78]. Moreover, silver-coated catheters effectively reduced urinary tract infection (UTI) up to 20% [79] and it was possible to reduce the risk of UTI significantly by using silver alloy-coated catheters [80]. Finally, it may be concluded that silver associated coating on urinary catheters are advantageous as well as it will eventually lower the cost.

Silver ions from associated coatings are able to reduce or prevent colonization and subsequent biofilm formation on the surface of the catheter and therefore associated infections can be reduced. In the case of silver nanoparticles, the release of silver ions from these nanoparticles is precisely controlled and thus it works more effectively [81, 82]. Therefore, it is expected that in the next few years the use of smart coating containing silver nanoparticles will be adopted for health applications. However, more vigorous research work is required to establish the efficacy of silver coating as well as to invent a proper way of application of such type of materials. During such research, several aspects related to infection should be taken care of e.g. bloodstream infection, urinary tract infection, ventriculomeningitis, morbidity, hospitalization time, economic costs etc. Hence, the real cause of infection should be identified first.

2.2.2.3 Working principle, limitations and side effect of silver coating

Metallic silver is less effective as the coated surface is strongly inhibited by protein adsorption on the surface of silver [64, 67]. However, the activity of silver nanoparticle depends on the size and shape of the particles [83-85]. In a study of Ho et al. [87], oxidized and reduced nanoparticles were produced under controlled conditions. A tendency of agglomeration of the commercially available silver nanoparticle is the main limitation for the use of this material as a surface coating material. One of the most influencing parameters is an available surface area for releasing Ag^+ ions, therefore, smaller and irregular nanoparticles show the highest antibiotic activity [86, 87].

Studies showed that nanoparticles bound themselves with cells and migrated into cells, damaging proteins, genetic material and membranes, leading to cell death [88-90]. The silver nanoparticles interact with bioactive components that transfer to the cell via a direct or indirect method and after that, the uptake of silver particle releases silver ions inside the cell. That silver ions will bind the bacterial cell membrane and proteins therefore, causing the destruction of the cell. Moreover, the intracellular silver nanoparticles cause damage to proteins and nucleic acids inside bacteria. In another way, the silver ions are directly bound to the cell membrane therefore, it causes the formation of perforation on the membrane through which the cellular compounds come out which ensures the cell death of bacteria.

Like other antibiotics, the long-term uses of silver or silver nanoparticles cause the development of silver-resistant bacteria [91, 92]. However, the important thing is that the growth of silver-resistant bacteria is lesser than the growth of antibiotic resistant bacteria. This happens because the silver attacks a number of cellular processes as well as the membrane integrity, whereas antibiotics target one specific process or enzyme. To protect the antibacterial activity of antibiotic the bacteria are mutated in such a way that the antibiotic is metabolized by the cell or removed from the cell. Sometimes the targeted enzyme by the antibiotic can be muted by the bacteria. On the other hand, some silver resistance genes which have been identified are responsible for pumping out silver from the cell [93] and the morphology of membrane changes in such way that silver affinity of the membrane is lowered down.

Potential side effects have been observed during extended use of silver and silver solutions. [94]. The exposure of silver particles under the skin causes the formation of grayish tinge [95-97]. The long-term use of silver nitrate or silver colloid solutions may affect the skin around the eye and discolored patches may be observed on the eyes. Such type of ophthalmological condition is called argyrosis [92] and that was caused by silver nitrate coated soft contact lenses [98]. The silver particles deposited on the eye and formed a blue-gray spot. Sometimes, the discoloration of scars and burns can be caused due to the use of silver-containing dressings [99]. Moreover, Silver could be accumulated in the liver and kidney with help of thiol-rich proteins like glutathione. Recent research showed that agyrosis could occur in the kidney of the human body [87]. Therefore, the controlled use of silver and silver nanoparticles may lead to increase the antibacterial properties of any surface.

3. Self-healing coatings: aspects, methods and applications

Polymer-based nonmetallic and metallic coatings are generally used for metal surfaces to protect those materials from the harmful environment. Recently surface coating materials having self-healing properties are frequently used and become the subject of research interest in the field of surface coatings. This chapter highlights the different types of self-healing coating, their various applications on different surfaces and different methods of applications.

3.1. Self-healing coating using polymers

For a long time polymers are being used as coating materials to protect the surfaces from harmful effects. Those polymers can be easily modified and have self-repairing properties. Therefore, those polymers and their composite have been used as self-healing coatings [100-114]. In several kinds of literature different type of self-healing coatings have been used so far, the details of which have been given in Table 1.

Table 1   Different type of self-healing coating on different metal surface.

Polymer	Nature of Surface	References
Polyaniline–molybdate (PANI–MoO$_4^{2-}$)	Iron	[100]
Dodecylbenzenesulfonicacid-doped polyaniline nano-particles (n-PANI(DBSA)) and neat epoxy ester	Iron	[101]
Polyaniline and silica sol–gel (PANI/sol–gel)	Aluminium	[102]
Polypyrrole doped with heteropolyanions of PMO$_{12}$O$_{40}^{3-}$ and HPO$_4^{2-}$	Iron	[103]
Poly (HFBMA-co-ITEGMA) Copolymer	Aluminium alloy	[104]

Poly(vinylidene fluoride) (PVDF)	Magnesium alloys	[105]
Ruthenium Grubbs' Catalyst (RGC) nanoparticles in 5-ethylidene-2-norbornene (5E2N) matrix	Not mentioned	[106]
Superabsorbent polymer ((SAP), known under the trade name 'AQALIC CS-7S', Nippon Shokubai Co. Ltd.), and a vinyl ester polymer (VEP)	Cold-rolled steel	[107]
Vinyl ester polymer with rutile	Aluminum alloy 5083	[108]
Polyelectrolytes, for example, poly(allylamine hydro-chloride) (PAH), poly(ethyleneimine) (PEI) with 2-(benzothazol-2-ylsulfanyl)succinic acid (BYS)	Aluminium alloy	[109]
Polystyrene sulphonate ion-exchange resin	Steel alloy	[110]
Porous polymer based on cellulose (CAF)	Cold-rolled steel plate	[111]
Polymer containing Chitosan	Aluminium alloy	[112]
Metallo-supramolecular gels containing 2,6-bis(1,2,3-trizol-4-yl)pyridine (BTP)	Metal surface	[113]
Epoxy materials based on maleimide chemistry	Not mentioned	[114]
Epoxy vinyl ester	Not mentioned	[115]
Epoxy with 15 wt% Poly(o-phenylenediamine) (PoPD)	Mild steel	[116]
Chitosan film doped with cerium ions	Aluminium alloy	[117]

Different research groups have followed different mechanisms. Among those conductive polymers like polyaniline (PANI), oxidize the steel surface and therefore the surface becomes passive. On the other hand, polypyrrole, which has an ion-exchange property, have been modified with $PMo_{12}O_{40}^{3-}$ and HPO_4^{2-} ions. Now the tetraoxomolybdate ions (MoO_4^{2-}) react with iron to have self-healing property on the coated surface. During the reaction iron molybdate has formed which significantly reduces substrate digestion. MoO_4^{2-} ions form during hydrolysis reaction [103], which can be represented as follows.

$$PMo_{12}O_{40}{}^{3-} + 12H_2O = 12\,MoO_4{}^{2-} + HPO_4{}^{2-} + 23H^+ \tag{1}$$

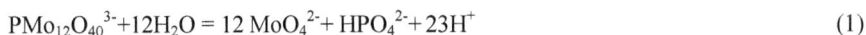

Polyaniline are conductive polymers and such polymers can displace the electro-active interface from its original location. PANI–MoO$_4{}^{2-}$ works as an oxidizer, therefore, it offers anodic galvanic protection to the coated surface from external corrosions [100]. A similar type observation was reported for polypyrrole (PPy) coatings enriched with PMo$_{12}$O$_{40}{}^{3-}$ and HPO$_4{}^{2-}$ ions. Having a self-cleaning property esters resin with n-PANI(DBSA) releases DBSA$^-$ anions, which reacts with iron cations to form an additional layer at the end of the surface coating [101]. To protect aluminum alloys, the surface can be coated with the PANI/sol–gel coating. Such type of coating has a self-repair mechanism with help of a redox reaction for the reduction of corrosion. During the redox reaction phase of polyaniline transforms from the emeraldine base form to the leucoemeraldine form [102]. Moreover, in the case of aluminum alloy 5083, vinylester polymer (VEP) coatings containing TiO$_2$ particles can be used to protect the alloy from corrosion. The bisphenol A (BPA) is a component of VEP and in reality BPA works as self-healing material. When the coated surface is damaged by any other means, VEP coating immediately releases BPA, which reacts with the aluminum and forms a passive layer on the open place of the alloy surface [108]. The TiO$_2$ particles preserve BPA inside the polymer.

AQALIC CS-7S polymer has water absorbing and swelling property and due to this property, less amount of oxygen can diffuse to the surface of coated cold-rolled steel [107]. The self-healing property of epoxy resins based coatings containing maleimides (e.g. 1, 1(methylenedi-4,1-phenyl) bismaleide or N,N'-(1,3-phenylenedi) maleimide)) and thiols (e.g. penta- erythritol(3-mercaptopropionate)) works in two different ways in the crack area. In the first case, self-repairing takes place at the crack interface while in the other case self-repairing takes place in the bulk area of the crack [115]. Billiet et al. [114] have explained that possibly maleimides reaction with the residual amines present in the epoxy matrix or with thiols. In the self-repairing reaction tertiary amines works as a catalyst and that is introduced into the polymer.

Epoxy resin with poly(o-phenylenediamine) (PoPD) nanotubes was used as a pigment and that was applied using a brushing technique onto mild steel surfaces. Sometime the coating can be removed or damaged and then the metal surface can be exposed to a corrosive environment. Then, the PoPD nanotubes cause the passivation of the uncovered surface to protect the material. The electrochemical oxidation reaction takes palace on that damaged surface during the reaction with help of atmospheric oxygen or the oxygen dissolved in the substrate works as the oxidizer, which oxidizes the metal surface. At the same time, the reduction of PoPD takes place [116].

The mixture of 5E2N and RGC nanoparticles is used as self-healing coating and ring opening metathesis polymerization (ROMP) of olefins is observed during the execution of self-healing property [106]. The rate of reaction is stimulated by thermal stimulation and the Grubbs catalyst compound (RGC), metallorganic compounds. However, there are some limitations for the the use of RGC i.e. it is very costly, toxic and not stable. Moreover, the prolonged exposure to oxygen, moisture and the amine-curing agent causes the lost of its property and activity [114].

Fluoropolymers have good water repellency property and can be used to protect magnesium alloys. Poly(vinylidene fluoride) coating reacts with magnesium hydroxide, generated during corrosion and forms passive magnesium fluoride on the surface of the magnesium alloy [105].

During the coating of poly(HFBMA-co-ITEGMA) copolymer on aluminum alloys, two layers are formed one is HFBMA, which behaves as a water barrier, is located on the solid–air interface and the second one is TEGMA, which is located on the solid–substrate interface. TEGMA has strong polarity and zipper-like mechanisms followed by such type of coating. The isocyanate groups react with water present in the air and form stable urea cross-links [104].

The biopolymers containing chitosan can be used as a coating material to protect aluminum alloys [112, 117]. The additive property of $Ce(NO_3)_3$ can improve the self-repair properties of such polymer as $Ce(NO_3)_3$ works as a corrosion inhibitor for aluminum alloys. At the damaged surface chitosan matrix releases Ce^{3+} cations and a passive layer of cerium hydroxides is formed on the damaged area.

Sometime multilayer coatings of polyelectrolyte are very effective for aluminum and associated alloys [118]. The working principle of such polyelectrolyte is different from the other polymeric coatings. First, it maintains such pH so that corrosive environment cannot affect aluminum. Moreover, polyelectrolyte can prevent any type of mechanical damages on the coated surface. The presence of 8-hydroxy-quinoline (8HQ) improves the quality of polyelectrolyte self-healing coating as this compound forms chelate of aluminum, which does not absorb chloride ions. The coating of strong negative polyelectrolyte–poly(styrene sulphonate) (PSS) along with weak positive polyelectrolyte poly(ethyleneimine) (PEI) forms a strong self-healing coating [119] and moreover coatings consisting of PEI and poly(allylamine hydrochloride) and 2-(benzothiazol-2-ylsulfanyl) succinic acid have self-repair properties [109].

Silyl ester (octyldimethyl- silyloleate), 8-hydroxyquinoline, calcium(II) and zinc(II) exchanged pigments, phenylphosphonic acid and sodium benzoate are introduced in their corresponding self-healing polymer matrix to enhance the performance as that material has good self-repair properties [110, 111, 120–122]. However, there are some limitations of polymer coatings where inhibitors are directly added to the matrix. Sometime the self-healing property does not work due to degradation of the coating's integrity, insufficient release or absence of inhibitors [123].

3.2 Self-healing coatings containing micro- or nano-capsules

Several research works have been carried out on encapsulation of inhibitors to enhance the performances of inhibitors as well as self-healing coatings. The studies showed that coating containing encapsulated inhibitor showed better performance than the coatings where the inhibitor is added directly [124]. The uncontrolled release of such inhibitors from the polymeric matrix is replaced by controlled release from micro- or nanocapsules. Therefore, the major advantage of encapsulation of inhibitors is to control the use of the self-repair properties. There are several parameters, which make changes in capsules and promote to release inhibitors. Those are generally activation with the light of different wavelengths, a change in the pH, a thermal change and magnetic, chemical, electrical and biological induction [125]. The self-repair action of coatings depends on the release of active molecules from with micro/nanocapsules and different types of release like desorption controlled release, pH controlled release, the ion-exchange control of release and release under mechanical rupture are observed [126].

The formation of micro/nano-capsules can be done by emulsification, layer-by-layer (LBL) assembly, coacervation and internal phase separation [125, 127]. The several materials used for encapsulation are poly(urea–formaldehyde) (PUF) [128–131], melamine–urea–formaldehyde (MUF) [131, 132], phenol–formaldehyde [133], epoxy resin [134, 135], methylene diphenyl diisocyanate (MDI) [136], poly(methyl methacrylate) (PMMA) [137, 138], polystyrene [128, 137], poly(allylamine) [139], polyvinyl alcohol (PVA) [139, 140], polyurethane [136], polyphenol, amphiphilic block copolymers (polypyrrolidones, poly(ethyleneoxide)) [127], poly(caprolactone)) [127, 128], cerium molybdate [141], zinc & aluminium nitrate [142,143], silica [123,128,144–146], silver [138], gold [147], copper(II) sulphide [147], c-AlO(OH) [147], tin(IV) oxide [147], Titanium oxides [138, 148, 149], $CaCO_3$ microbeads [150], halloysite nanotubes and boehmite nanoparticles [153–155], etc. There are many types of mechanisms available for the preparation of micro/nano-capsules and according to those the shape and size of the capsules vary. The speed of agitation is the main size controlling parameter and higher speed results in smaller size capsules. [136,151]. In general, the capsules are spherical but tubular capsules can be generated in case of SiO_2 and PMMA [152]. Some time to get better performance polymer capsules are coated and annealed at $500^\circ C$ with inorganic oxide like SiO_2 or TiO_2 [120,144]. The major inhibiters, which are present inside the capsules are cerium cations [154,156–158], 8-hydroxyquinoline [144,153], benzotriazole [153, 158], 2-mercaptobenzothiazole (MBT) [141, 144, 145, 159, 160], mercaptobenzimidazole (MBI) [159], dicyclopentadiene (DCPD), alkoxysilanes [161], silyl ester [120], linseed oil [133, 140, 162], hexamethylene diisocyanate (HDI) [136], triethanolamine (TEA) [137], monomers and catalysts for ROMP [163], magnesium ions [164], chromium(III) and cerium(IV) oxides [130] and cerium(III) chloride [156, 165]. The important inhibitors and along with corresponding encapsulation materials are given in Table 2.

Table 2   Different types of indictors and their corresponding encapsulation materials.

Encapsulation material	Inhibitor	Reference
Formaldehyde	Linseed oil	[133]
Methylene diphenyl diisocyanate (MDI)	Hexamethylene diisocyanate (HDI)	[136]
Poly(methyl methacrylate) (PMMA)	Triethanolamine (TEA)	[137]
Polystyrene	Triethanolamine (TEA)	[137]
Polyvinyl alcohol (PVA)	Linseed oil	[140]
Polyurethane	Hexamethylene diisocyanate (HDI)	[136]
Cerium molybdate	2-mercaptobenzothiazole (MBT)	[141]
Zinc and aluminium nitrate	Vanadate, phosphate, and 2-mercaptobenzothiazolate	[142,143]
Silica	8-hydroxyquinoline, 2-mercaptobenzothiazole (MBT)	[144]
Silica	2-mercaptobenzothiazole (MBT)	[145]
Gold, copper(II) sulphide, c-alo(OH), tin(IV) oxide	KSCN and phenylalanine	[147]

Titania oxides	8-hydroxyquinoline	[149]
CaCO₃ microbeads	Cerium nitrate, salicylaldoxime and 2,5-dimercapto-1,3,4-thiadiazolate	[150]

In all the studies, it has been observed that first encapsulation of the active material is completed then micro/nanocontainers dispersed in the coating matrix. Several techniques are adopted to fill the capsules and mainly some driving forces are applied for filling purpose and encapsulation can be done by agitation of capsules and fillers, under reduced pressure, the soaking of capsules in inhibitor solution and applying the LBL deposition procedure [123,150,152,166]. In the case of soaking encapsulation, different mechanisms come under consideration like the ion-exchange mechanism and adsorption inhibitors on the porous surface of the capsules. The CaCO₃ microbeads have been used as containers for the adsorption cerium ions and organic inhibitors [150]. Sometimes micro/nanocapsules are coated with different materials, mainly polymeric coatings [131, 150, 164], in few cases sol–gel coatings [123,168], and metallic coatings, e.g. electrochemical nickel coatings [169]. Several parameters like concentration and the position of capsules containing inhibitors in the coating matrix have a significant effect on the self-healing property. The concentration of self-repair should be maintained in such a way that inhibitor ensures sufficient degree of protection and the size and shape of the micro/nanocapsules should be maintained in a proper way so that they can cover themselves inside the coating. Studies have shown that the location of capsules close to the metal surface provides better active protection against corrosion than the capsules present in the coating matrix away from the metal surface [123].

As above-mentioned, there are several advantages of self-healing coatings containing micro- or nanocapsules but like every process, there must be some disadvantages. In most cases, self-healing takes palace only after the first damage to the coating. Moreover, the incorporation of micro- or nanocapsules adversely affects the adhesion property of the coating to the substrate [127,142,163,170,171]. To get the maximum advantage of coating capsules of inhibitor, the material of micro/nano container material should be matched with the coating material e.g. silica nanocontainers does not suit with polymer-based coatings, as their mechanical properties are different. Therefore, in such cases, polymer nano-containers are more suitable rather than silica [163]. To eliminate unnecessary circumstances in using micro/nanocontainer, gel particles with micron size (microgel) can be used and this is very simple indeed. The polyuria microgel particles with 2-methylbenzothiazole (MeBT) works as an inhibitor of a self-healing coating system and thus entrap in the solid matrix [172].

3.3   Self-healing coating using hybrid oxide

Several techniques have been adapted to form self-healing hybrid oxide coatings like sol–gel technique [173–177], electrodeposition [178], and plasma techniques [179]. Commercially available alcoholates or salts of several metals and silicon are generally used as the substrate. Different oxides are formed in the sol–gel technique and those oxides e.g. SiO_2, ZrO_2, TiO_2 contribute various protective properties to the coating materials [173,174,176,180–191]. Mainly steel or aluminum alloys are protected by using such coatings and this type of coating is a better replacement of the chromate-based coating. Lanthanides ions, benzotriazoles, 8-hydroxy- quinoline, propargyl alcohol or red mud (waste from the production of aluminum by the Bayer method, consisting of iron and titanium oxides and silicon compounds) are the major inhibitors which are commonly introduced to hybrid oxide coating to obtain self-healing properties [177,180,186–195]. In the hybrid oxide coating, inorganic groups are attached with native oxide film on the metal surface by forming covalent bonds whereas the organic groups are attached to the painting systems. Therefore, a molecular network is established inside the coated surface and such type of molecular attachments improve the adhesive properties of the coating material as well as the performance of the coated metallic surface. Not only those inorganic oxides like silica are added to the hybrid coating but some organic compounds such as methacrylates, alkylsilanes and poly(ethylene imine) (PEI) are also incorporated. Those organic fractions improve the ductility of the system. For example, PEI fulfills different requirements one it behaves, as a cross-linking agent for hybrid coating formation and the second one is it solubilizes the organic corrosion inhibitor. The cerium incorporated into montmorillonite nanoparticles, MBT or 2-mercaptobenzimidazole (MBI) to improve anti-corrosion resistance of hybrid coating [187,194]. Sometimes, the lanthanides ions of self-healing coating diffuse to substrate surface and form an insoluble layer of oxides or hydroxides [173,174,180–182]. The inhibition property of Ce(III), Ce(IV) ions is quite similar to that of Cr(VI) ion thus the former two ions can be used as a replacement of Cr(VI) [196]. However, the presence of lanthanide leads to weakening the surface coating as loss of the inhibitor takes place through the porous structure of the lanthanide. The stability of surface coating containing red mud increased due to the formation of aluminum oxide after annealing at a temperature of about $900^{\circ}C$ [188]. Madani et al. [189] reported that Benzotriazole (BTA), an anodic inhibitor, adsorbed on the metallic surface but Yasakau et al. [192] revealed that the presence of BTA in the sol–gel coating negatively affected the anti- corrosion properties of the coated surface.

Propargyl alcohol and 8HQ can positively influence the barrier properties of the coating [190,192].

Sometimes smart coating can be done by direct formation of a porous layer of TiO_2 nanoparticles and after that, this layer is covered with a sol–gel coating consisting of zirconium oxide and organosiloxane [148,158]. The nanostructured TiO_2 is used as coating material as it is resistant to chemical reaction with other compounds and it has photo-electrochemical properties. Due to the presence of this unique property, such type of nanoparticles can be used to build solar cells and for the photogenerated cathodic protection of metals under ultraviolet (UV) illumination. Under UV illumination, electrons are transferred from the valence to conduction band of TiO_2 thus the metal surface becomes negatively charged and ultimately the metal gets its thermodynamic stability. Therefore, the metal surface remains unaffected by the corrosion. To ensure active protection under a limited access of UV irradiation, the coating material is doped with chromium(III) ions which help to produce electron form TiO_2 at the access of energy. Chromium(III) ions work as a promoter of the photocatalytic activity of TiO_2 nanoparticles [197].

The concentration of silica is a very important property of silica-based surface coating materials and that value should be optimized. The structure of silica is porous, therefore, with an increase of silica concentration the porosity of the coating material also increases. Hence, to improve the performance of silica based coatings some material can be added so that silica can establish adhesive bonding with a metal surface like Si–O–M (M represents a metal constituent) [198].

3.4   Other smart self-healing coatings

The self-healing conversion coating was successfully used to coat galvanized steel and silicate coating of SiO_2 and Na_2O mixture was deposited on the metal surface. During coating operation, the silicate anions react with the metallic substrate and form a new compound, which contained Zn, O and Si, acted as anti-corrosion material [201]. The toxic chromate based conversion coatings can be replaced vanadia based coatings. During the self-healing coating process, a passive oxide layer of vanadia has been formed [199, 200, 202]. Conversion coatings based on potassium stannate has formed corrosion resistant tin oxide-rich magnesium hydroxide layer, which prevents oxygen diffusion to the magnesium alloy surface [203]. Yabuki and Kaneda [204] have used fluoro-organic compounds having different terminal groups (COOH, OH, COF) for surface coating on metallic zinc and such type of coating forms a barrier film on damaged. Dipped coating method has been performed using casein solution with 1 wt.% of TiO_2. In this case, self-healing property is totally pH dependent and with an increase of pH on the damaged surface causes the digestion of the alloy. After that, casein releases from the coating, which is responsible for the diffusion of TiO_2 particles to the damaged place and a passive barrier layer is formed on the exposed surface. As a result, magnesium alloy is protected from any type of corrosion [205]. Ceramic materials containing TiC/Al_2O_3 and Ti_2AlC are used for self-healing coatings and with an increase of temperature, a self-healing property is activated. At high temperature, the oxidation takes place by forming TiO_2 and Al_2O_3, which fill up the defects in the surface coating [206,207].

From the above discussion, it can be concluded that self-healing coatings make the surface more sustainable in an unfavorable environment and thus the lifetime of a coated surface is far better than the uncoated surface. Therefore, the overall maintenance cost of the uncoated surface can be reduced by applying such type of coating.

4.   Anti-fogging coatings: aspects, methods and applications

The anti-fogging coatings have several advantages and the nature of these coatings is hydrophilic. Due to such nature, anti-fog film causes moisture to spread into an even layer by wetting the surface so it will not form droplets, which obscure vision. In the current scenario, only a few numbers of literature are available regarding such coatings. However, some studies have elaborately described various techniques to improve the surface properties of different optical devices [208-212]. The presence of fog on the surface of medical and analytical instruments can reduce the performance and efficiency of such instruments. Moreover, the surface which is born to fog formation cause problem to the light transmission and therefore optical efficiency of such instrument like eyeglasses, goggles, face shields, binoculars, etc. get reduced [4]. During the fog generation, water vapor condenses onto surfaces and form discrete water droplets. Light diffuses through the droplets and after that, the light is scattered. By applying the anti-fogging coating, the surface energy is increased and therefore hydrophilic property of the coated surface is improved. Hence, the condensation of water vapor takes place when it comes in the contact with such hydrophilic surface and produces a continuous, transparent thin film [211].

4.1.   Anti-fogging coating using polymers and monomers

In the last few years, several researchers have placed their effort to generate an effective anti-fogging coating by applying a thin film of different materials [213-221]. Among them, most researchers have obtained an anti-fogging hydrophilic surface by depositing a thin film of polymers or monomers containing hydrophilic functionalities, such as hydroxyl (OH) or carboxyl groups (COOH, COOR). In general, anti-fogging coatings have been employed using poly(ethylene glycol) [222], poly (vinyl acetate) [223,226-

228] poly(vinyl alcohol) [217-219,222-225] cellulose ester or cellulose ether [222, 223], acrylic resin with pending OH or COOH groups [222,229,230] glycidyl derivatives [231,232] or poly(vinyl pyrrolidone) [214, 215]. The details of different types of polymers, which have been used to enhance the hydrophilic properties, have been listed in Table 3. Due to the presence of the hydrophilic group, those polymers can dissolve by absorbing water molecule during wetting. Therefore, to enhance the stability of the coating several techniques have been adopted in recent past.

Table 3   Different components which are used for anti-fogging coating.

Polymers	Surface	Reference
Poly(ethylene glycol)	Lenses	[222]
Poly (vinyl acetate)	Glass or plastic substrate	[223]
Poly(vinyl alcohol) (PVA)	Optical Surface	[225]
Cellulose ester or cellulose ether	Lenses, glass or plastic substrate	[222, 223]
Acrylic resin	Lenses, glass surfaces	[222, 229, 230]
Glycidyl derivatives	Alloys	[231]
Poly(vinyl pyrrolidone)	Optical surface	[215]

The reticulation agents can be used to enhance the stability of the coating as this type of agents establish a cross-linked polymeric network and such type of agents may be organic or inorganic compounds or a mixture of both e.g. polyisocyanate [214, 222], glycol derivatives [229], acrylamide and epoxy resins [222,233] aluminium derivatives [223], or a combination of formaldehyde and zirconium nitrate [219]. The cross-linking process with help of reticulation agent can be carried out under either UV irradiation [222,229, 231,232] or heating [217,218, 223]. This cross-linked network prevents coating material from dissolving in the water film. In 2011, Chevallier et al. [233] followed three steps to apply an anti-fogging coating on fused silica. Before the plasma treatment, the glass surface was activated using a piranha solution, which oxidized the surface. This technique resulted in the grafting of 1.2% of amino groups on the fused silica surface. After that poly(ethylene-alt-maleic anhydride) (PEMA) coating was employed on the modified silica using spinning action and then it cured under $95^{\circ}C$ for overnight. It was again coated with PVA and this coating process was used for both single layer and multiple layers of PEMA/PVA coating to generate the anti-fogging surface.

4.2 Applications of anti-fogging coating

As mentioned earlier, that anti-fogging surface coating is mainly applied to enhance visual clarity. Moreover, this type of coating can reduce several accidents happened in our daily lives and it elongates the lifetime of the coated surface. The major applications are as follows:

- Automotive & Transit: windows, side mirrors, windshields & screens.
- Bathroom: residential and commercial mirrors, windows, shower and steam room doors.
- Commercial freezer display and deli cases.
- Electronics: HUD, LED, and outdoor digital billboard display.
- Optical Uses: eyeglasses, goggles, face shields, binoculars, microscopes, medical instruments.

There are several advantages regarding the use of anti-fogging coatings.

5. Conclusion

The details of applications of different types of smart coatings and methods related to surface coatings have been elaborately described in this chapter. It has been concluded that smart coatings have tremendous applications in our daily life and can be used for the betterment of our life as well. However, more commercialization of surface coatings is required to avail the benefits of such technologies for the common people.

References

[1] R. Gradinger, R. Graf, J. Grifka, J. Löhr, Das infizierte implantat, Der. Orthopäde, 3 (2008) 257-269.

[2] M, Hellmann, S.D. Mehta, D.M. Bishai, S.C. Mears, J.M. Zenilman. The estimated magnitude and direct hospital costs of prosthetic joint infections in the United States, 1997 to 2004, J. Athroplast., 25 (2010) 766-771. https://doi.org/10.1016/j.arth.2009.05.025

[3] M.F. Sampredo, R. Patel, Infections associated with long-term prosthetic devices, Infect. Dis. Clin. N. Am., 21 (2007), 785-819. https://doi.org/10.1016/j.idc.2007.07.001

[4]   A.J. Tokarczyk, S.B. Greenberg, J.S. Vender, Death, dollars, and diligence: Prevention of catheter-related bloodstream infections must persist, Crit. Care Med., 37 (2009) 2320-2321. https://doi.org/10.1097/CCM.0b013e3181a9efa9

[5]   C. Von eiff, B. Jansen, W. Kohnen, K. Becker, Infections associated with medical devices, Drugs, 65 (2005) 179-214. https://doi.org/10.2165/00003495-200565020-00003

[6]   J.D. Turnidge, D. Kotsanas, W. Munckhof, S. Roberts, C.M. Bennett, G.R. Nimmo, G.W. Coombs, R.J. Murray, B. Howden, P.D.R. Johnson, K. Dowling, Staphylococcus aureus bacterimia: a major cause of mortality in Australia and New Zealand, Med. J. Aust., 191 (2009) 368-373.

[7]   M.A. Olsen, S. Chu-Ongsakul, K.E. Brandt, J.R. Dietz, J. Mayfeld, V.J. Fraser, Horpital-associated costs due to surgical site infection after breast surgery, Arch. Surg., 143 (2008) 53-60. https://doi.org/10.1001/archsurg.2007.11

[8]   S. Noimark, C.W. Dunnill, M. Wilson, I.P. Parkin, The role of surfaces in catheter-associated infections, Chem. Soc. Rev., 38 (2009) 3435-3448. https://doi.org/10.1039/b908260c

[9]   B.W. Trautner, R.O. Darouiche, Catheter-associated infections, Arch. Intern. Med.,164 (2004) 842-850. https://doi.org/10.1001/archinte.164.8.842

[10]   A. David, D.C. Risitano, G. Mazzeo, L. Sinardi, F.S. Venuti, A.U. Sinardi, Central venous catheters and infections, Minerva. Anestesiol., 71 (2005) 561-564.

[11]   K. Halton, N. Graves, Economic evaluation and catheter-related bloodstream infections, Emerg. Infect. Dis., 13 (2007) 815-823. https://doi.org/10.3201/eid1306.070048

[12]   E.N. Prencevich, D. Pittet, Preventing catheter-related bloodstream infections, J. Am. Med. Assoc., 301(2009) 1285-1287. https://doi.org/10.1001/jama.2009.420

[13]   W. Zingg, A. Imhof, M. Maggiorini, R. Stocker, E. Keller, C. Ruef., Impact of a prevention strategy targeting hand hygiene and catheter care on the incidence of catheter-related bloodstream infections, Crit. Care. Med., 37 (2009) 2167-2173. https://doi.org/10.1097/CCM.0b013e3181a02d8f

[14]   A.J. Kallen, P.R. Patel, N.P. O'Grady, Preventing catheter-related bloodstream infections outside the intensive care unit: Expanding to new settings, Healthcare Epidem., 51 (2010) 335-341. https://doi.org/10.1086/653942

[15]   J.M. Walz, S.G. Memtsoudis, S.O. Heard, Prevention of central venous catheter bloodstream infections, J. Intensive. Care. Med., 25 (1997) 131-138. https://doi.org/10.1177/0885066609358952

[16]   B. A˝ıssa, D. Therriault, E. Haddad, W. Jamroz, Self-Healing Materials Systems: Overview of Major Approaches and Recent Developed Technologies, Adv. Mater. Sci. Eng., 2012 (2012) 1-17. https://doi.org/10.1155/2012/854203

[17]   V.D. Rosenthal, D.G. Maki, N. Graves, The international nosocomical infection control consortium (INICC): Goals and objectives, description of surveillance methods, and operational activities, Am. J. Infect. Control., 36 (2008) e1-e12. https://doi.org/10.1016/j.ajic.2008.06.003

[18]   G.M.L. Bearman, C. Munro, C. N. Sessler, R.P. Wenzel, Infection control and the prevention of nosocomical infections in the intensive care unit, Sem. Resp. Crit. Care. Med., 27 (2006) 310-324. https://doi.org/10.1055/s-2006-945534

[19]   W.T. Youg, How to respond to changes in the regulation of the ethylene-oxide sterilization process, Med. Dev. Technol., 17 (2006) 12-15.

[20]   P. Thevenot, W. Hu, L. Tang, Surface chemistry influences implant biocompatibility, Curr. Top. Med. Chem., 8 (2008) 270-280. https://doi.org/10.2174/156802608783790901

[21]   H. Stutz, Protein attachment onto silica surfaces—A survey of molecular fundamentals, resulting effects and novel preventive strategies in CE, Electrophoresis, 30 (2009) 2032-2061. https://doi.org/10.1002/elps.200900015

[22]   F. Ganazzoli, G. Raffaini, Computer simulation of polypeptide adsorption on model biomaterials, Phys. Chem. Chem. Phys., 7 (2005) 365-3663. https://doi.org/10.1039/b506813d

[23]   D. Pavithra, M. Doble. Biofilm formation, bacterial adhesion and host response on polymeric implants—issues and prevention, Biomed. Mater., 3 (2008) 1-13. https://doi.org/10.1088/1748-6041/3/3/034003

[24]   M. Cloutier, D. Mantovani, F.Rosei, Antibacterial coatings: challenges, perspectives, and opportunities, Trends Biotechnol., 33 (2015) 637-652. https://doi.org/10.1016/j.tibtech.2015.09.002

[25]   L. Hall-Stoodley, P. Stoodley, Evolving concepts in biofilm infections, Cell. Microbiol., 11 (2009) 1034-1043. https://doi.org/10.1111/j.1462-5822.2009.01323.x

[26]   C. Sousa, P. Teixeira, R. Oliveira, Influence of surface properties on the adhesion of Staphylococcus epidemidis to acrylic and silicone, Int. J. Biomater., 2009 (2009) 1-9. https://doi.org/10.1155/2009/718017

[27]   M. Katsikogianni, I. Spiliopoulou, D.P. Dowling, Y.F. Missirlis, Adhesion of slime producing Staphylococcus epidermidis strains to PVC and diamond-like carbon/silver/fluorinated coatings, J. Mater. Sci., 17 (2006) 679-689. https://doi.org/10.1007/s10856-006-9678-8

[28]   A. Almaguer-Flores, L.A. Ximenez-Fyvie, S.E. Rodil. Oral bacterial adhesion on amorphous carbon and titanium films: Effect of surface roughness and culture media, J. Biomed. Mater. Sci. B. Appl. Biomater., 92B (2010) 196-204. https://doi.org/10.1002/jbm.b.31506

[29]   J. Hoffmann, J. Groll, J. Heuts, H. Rong, D. Klee, G. Ziemer, M. Moeller, H.P. Wendel, Blood cell and plasma protein repellent properties of star-PEG-modified surfaces, J. Biomater. Sci. Polym. Ed., 17 (2006) 985-996. https://doi.org/10.1163/156856206778366059

[30]   I. Fundeanu, D. Klee, A.J. Schouten, H.J. Busscher, H.C. van der Mei, Solvent-free functionalization of silicone rubber and efficacy of PAAm brushes grafted from an amino-PPX layer against bacterial adhesion, Acta Biomater., 6 (2010) 4271-4276. https://doi.org/10.1016/j.actbio.2010.06.010

[31]   G. Cheng, H. Xue, Z. Zhang, S. Chen, S.A. Jiang, A switchable biocompatible polymer surface with self-sterilizing and nonfouling capabilities, Angew. Chem. Int. Ed., 47 (2008) 8831-8834. https://doi.org/10.1002/anie.200803570

[32]   W. Senaratne, L. Andruzzi, C.K. Ober, Self-assembled monolayers and polymer brushes in biotechnology: Current applications and future perspectives., Biomacromolecules., 6 (2005) 2427-2448. https://doi.org/10.1021/bm050180a

[33]   F. Schreiber, Structure and growth of self-assembling monolayers, Prog. Surf. Sci. 65 (2000) 151-256. https://doi.org/10.1016/S0079-6816(00)00024-1

[34]   J.P. Bearinger, S. Terrettaz, R. Michel, N. Tirelli, H. Vogel, M. Textor, J.A. Hubbell, Chemiadsorbed poly(propylene sulphide)-based copolymers resist biomolecular interactions, Nat. Mater., 2 (2003) 259-264. https://doi.org/10.1038/nmat851

[35]   X. Khoo, P. Hamilton, G.A. O'Toole, B.D. Snyder, D.J. Kenan, M.W. Grinstaff, Directed assembly of PEGylated-peptide coatings for infection-resistant titanium

metal, J. Am. Chem. Soc., 131 (2009) 10992-10997.
https://doi.org/10.1021/ja9020827

[36]   R.G.J.C. Heijkants, Nanotechnology delivers microcoatings, Med. Device.
Technol., 17 (2006) 14-16.

[37]   A.W. Bridges, A.J. Garcia, Anti-inflammatory polymeric coatings for implantable
biomaterials and devices, J. Diabet. Sci. Technol., 2 (2008) 984-994.
https://doi.org/10.1177/193229680800200628

[38]   A. Roosjen, H.C. van der Mei, H.J. Busscher, W. Norde, Microbial adhesion to
poly(ethyle oxide) brushes: Influence of polymer chain length and temperature,
Langmuir., 20 (2004) 10949-10955. https://doi.org/10.1021/la048469l

[39]   T. Thorsteinsson, T. Loftsson, M. Masson, Soft antibacterial agents, Curr. Med.
Chem., 10 (2003) 1129-1136. https://doi.org/10.2174/0929867033457520

[40]   H. Murata, R.R. Koepsel, K. Matyjaszewski, A.J. Russell, Permanent, non-
leaching antibacterial surfaces-2: How high density cationic surfaces kill bacterial
cells. Biomater., 28 (2007) 4870-4879.
https://doi.org/10.1016/j.biomaterials.2007.06.012

[41]   G. Cheng, H. Xue, G. Li, S. Jiang, Integrated antimicrobial and nonfouling
hydrogels to inhibit the growth of planktonic bacterial cells and keep the surface
clean, Langmuir., 26 (2010) 10425-10428. https://doi.org/10.1021/la101542m

[42]   L.B. Rawlinson, P.J. O'Brien, D.J. Brayden, High content analysis of cytotoxic
effects of pDMAEMA on human intestinal epithelial and monocytes cultures., J.
Contr. Rel., 146 (2010) 84-92. https://doi.org/10.1016/j.jconrel.2010.05.002

[43]   S. Venkataraman, Y. Zhang, L. Liu, Y. Yang, Design, synthesis and evaluation of
hemocompatible pegylated-antimcrobial polymers with well-controlled molecular
structures, Biomater., 31 (2010) 1751-1756.
https://doi.org/10.1016/j.biomaterials.2009.11.030

[44]   D.L. Fredell, Biological properties and applications of cationic surfactants. In
Cationic Surfactants, 1st ed.; Cross, J.; Singer, E.J.; Eds.; Marcel Dekker Inc.:
New York, NY, USA, (1990) 31-60.

[45]   T. Ravikumar, H. Murata, R.R. Koepsel, A.J. Russell, Surface-active antifungal
polyqyaternary amine., Biomacromol., 7 (2006) 2762-2769.
https://doi.org/10.1021/bm060476w

[46]   K. Hegstad, S. Langsrud, B.T. Lunestad, A.A. Scheie, M. Sunce, S.P.
Yazdankhah, Does the wide use of quaternary ammonium compounds enhance the

selection and spread of antimicrobial resistance and thus threaten our health? Microb. Drug. Resist., 16 (2010) 91-104. https://doi.org/10.1089/mdr.2009.0120

[47]   A. Fleming, On the antibacterial action of cultures of a Penicilium, with special reference to their use in the isolation of B. influenzae, Br. J. Exp. Pathol., 10 (1929) 226-236.

[48]   T. Jaeblon, Polymethylmethacrylate: Properties and contemporary uses in orthopaedics, J. Am. Acad. Otrhop. Surg., 18 (2010) 297-305. https://doi.org/10.5435/00124635-201005000-00006

[49]   J.C.J. Webb, R.F. Spencer, The role of polymethylmethacrylate bone cement in modern orthopaedic surgery, J. Bone. Joint. Surg., 89-B (2007) 851-857. https://doi.org/10.1302/0301-620X.89B7.19148

[50]   P.A. Norowski, J.D. Bumgardner, Biomaterial and antibiotic strategies for peri-implants, J. Biomed. Mater. Res. B. Appl. Biomater. 88B (2009) 530-543. https://doi.org/10.1002/jbm.b.31152

[51]   L. Zhao, P.K. Chu, Y. Zhang, Z. Wu, Antibacterial coatings on titanium implants, J. Biomed. Mater. Res. B. Appl. Biomater, 91B (2009) 470-480. https://doi.org/10.1002/jbm.b.31463

[52]   T. Kälicke, J. Schierholz, U. Schlegel, T.M. Frangen, M. Köller, G. Printzen, D. Seybold, S. Klöckner, G. Muhr, S. Arens, Effect on infection resistance of a local antiseptic and antibiotic coating on osteosynthesis implants: An in vitro and in vivo study, J. Orthop. Res., 24 (2006) 1622-1640. https://doi.org/10.1002/jor.20193

[53]   L.G. Harris, L. Mead, E. Müller-Oberländer, R.G. Richards, Bacteria and cell cytocompatibility studies on coated medical grade titanium surfaces, J. Biomed. Mater. Res., 78A (2006) 50-58. https://doi.org/10.1002/jbm.a.30611

[54]   M. Stigter, J. Bezemer, K. de Groot, P. Layrolle, Incorporation of different antibiotics into carbonated hydrozyapatite coatings on titanium implants, release and antibiotic effect, J. Contr. Rel., 99 (2004) 127-137. https://doi.org/10.1016/j.jconrel.2004.06.011

[55]   R.O. Darouiche, M.D. Mansouri, D. Zakarevicz, A. Alsharif, G.C. Landon, In vivo efficacy of antimicrobial-coated devices, J. Bone. Joint. Surg. Am., 89 (2007) 792-797.

[56]   M.E. Falagas, K. Fragoulis, I.A. Bliziotis, I. Chatzinikolaou, Rifampicin-impregnated central venous catheters: A meta-analysis of randomized controlled

trials, J. Antimicrob. Chemother., 59 (2007) 359-369.
https://doi.org/10.1093/jac/dkl522

[57] T. Hernandez-Richter, H.M. Schardey, F. Wittmann, S. Mayr, M. Schmitt-Sody, S. Blasenbrue, M.M. Heiss, C. Gabka, M.K. Angele, Rifampicin and triclosan but not silver is effective in preventing bacterial infection of vascular Dacron graft material, Eur. J. Vasc. Endovasc. Surg., 26 (2003) 550-557.
https://doi.org/10.1016/S1078-5884(03)00344-7

[58] K.A. Halton, D. Cook, D.L. Paterson, N. Safdar, N. Graves, Cost-effectiveness of a central venous catheter care bundle, PloS. ONE. (2010),
http://dx.doi.org/10.1371/journal.pone.0012815.
https://doi.org/10.1371/journal.pone.0012815

[59] B. Walder, D. Pittet, M.R. Tramer, Prevention of bloodstream infections with central venous catheters treated with anti-infective agents depends on catheter type and insertion time: Evidence from a meta-analysis, Infect. Control. Hosp. Epidemiol., 23 (2002) 748-756. https://doi.org/10.1086/502005

[60] H. Ceri, M.E. Olson, R.J. Turner, Needed, new paradigms in antibiotic development, Expert. Opin. Pharmacother., 11 (2010) 1233-1237.
https://doi.org/10.1517/14656561003724747

[61] A. Lohda, A.D. Furlan, H. Whyte, A.M. Moore, Prophylactic antibiotics in the prevention of catheter-associated bloodstream bacterial infection in preterm neonates: a systematic review, J. Perionatol., 28 (2008) 526-533.
https://doi.org/10.1038/jp.2008.31

[62] J.W. Alexander, History of the medical use of silver, Surg. Infect., 10 (2009) 289-292. https://doi.org/10.1089/sur.2008.9941

[63] M. Spear, Silver: an age-old treatment modality in modern times, Plast. Surg. Nurs., 30 (2010) 90-93. https://doi.org/10.1097/PSN.0b013e3181deea2e

[64] B.S. Atiyeh, M. Costagliola, S.N. Hayek, S.A. Dibo, Effect of silver on burn wound infection and healing: Review of the literature, Burns., 33 (2007) 139-148.
https://doi.org/10.1016/j.burns.2006.06.010

[65] S. Silver, L.T. Phung, G. Silver, Silver as biocides in burn and wound dressings and bacterial resistance to silver compounds, J. Ind. Microbiol. BioTechnol., 33 (2006) 627-634. https://doi.org/10.1007/s10295-006-0139-7

[66] H.J. Klasen, Historical review of the use of silver in the treatment of burns. I. early uses., Burns., 26 (2000) 117-130. https://doi.org/10.1016/S0305-4179(99)00108-4

[67] M. Rai, A. Yadav, A. Gade, Silver nanoparticles as a new generation of antimicrobials, Biotechnol. Adv., (2009) 76-83. https://doi.org/10.1016/j.biotechadv.2008.09.002

[68] T. Nakane, H. Gomyo, I. Sasaki, Y. Kimoto, N. Hanzawa, Y. Teshima, T. Namba, New antiaxillary odour deodorant made with antimicrobial Ag-eolite (silver-exchanged zeolite), Int. J. Cosmet. Sci., 28 (2006) 299-309. https://doi.org/10.1111/j.1467-2494.2006.00322.x

[69] J.L. Meakins, Silver and new technology: dressings and devices Surg. Infect., 10 (2009) 293-296. https://doi.org/10.1089/sur.2008.9942

[70] C. Dowsett, The use of silver-based dressing in wound care, Nurs. Stand., 19 (2004) 56-60. https://doi.org/10.7748/ns.19.7.56.s58

[71] M.N. Strom-Versloot, C.G. Vos, D.T. Ubbink, H. Vermeulen, Topical silver for preventing wound infection, Cochrane. Database. Syst. Rev., 3 (2010) 1-110. https://doi.org/10.1002/14651858.CD006478.pub2

[72] G. Gravante, R. Caruso, R. Sorge, F. Nicoli, P. Gentile, V. Cervelli, Nanocrystalline silver. Reconstr. Surg. Burns. 63 (2009) 201-205. https://doi.org/10.1097/SAP.0b013e3181893825

[73] G. Gravante, A. Montone, A retrospective analysis of ambulatory burn patients: Focus on wound dressings and healing times, Ann. Rev. Coll. Surg. Engl., 92 (2010) 118-123. https://doi.org/10.1308/003588410X12518836439001

[74] J. Fichtner, E. Güresir, V. Seifert, A. Raabe, Efficacy of silver-bearing external ventricular drainage catheters: A retrospective analysis, J. Neurosurg. 112 (2010) 840-846. https://doi.org/10.3171/2009.8.JNS091297

[75] D. Roe, B. Karandikar, N. Bonn-Savage, B. Gibbins, J.B. Roullet, Antimicrobial surface functionalization of plastic catheters by silver nanoparticles, J. Antimicrob. Chemther., 61 (2008) 869-876. https://doi.org/10.1093/jac/dkn034

[76] S.H. Hsu, H.J. Tseng, Y.C. Lin, The biocompatibility and antibacterial properties of waterborne polyurethane-silver nanocomposites, Biomater., 31 (2010) 6796-6808. https://doi.org/10.1016/j.biomaterials.2010.05.015

[77] J.R. Johnson, M.A. Kuskowski, T.J. de Wilt, Systematic review: Antimicrobial urinary catheters to prevent catheter-associated urinary tract infection in hospitalized patients, Ann. Intern. Med., 144 (2006) 116-126. https://doi.org/10.7326/0003-4819-144-2-200601170-00009

[78] B. Trautner, Management of catheter-associated urinary tract infection, Curr. Opin. Infect. Dis., 23 (2010) 76-82. https://doi.org/10.1097/QCO.0b013e328334dda8

[79] C. Seymour, Audit of catheter-associated UTI using silver alloy-coated Foley catheters., Br. J. Nurs., 15 (2006) 598-603. https://doi.org/10.12968/bjon.2006.15.11.21227

[80] D. Parker, L. Callan, J. Harwood, D.L. Thompson, M. Wilde, M. Gray, Nursing interventions to reduce the risk of catheter-associated urinary tract infection, J. Wound. Ostomy. Cont. Nurs., 36 (2009) 23-34. https://doi.org/10.1097/01.WON.0000345173.05376.3e

[81] J. Liu, D.A. Sonshine, S. Shervani, R.H. Hurt, Controlled release of biologically active silver from nanosilver surfaces, ACS. Nano, 4 (2010) 6903-6913. https://doi.org/10.1021/nn102272n

[82] P. Pallavicini, A. Taglietti, G. Dacarro, Y.A. Diaz-Fernandez, M. Galli, P. Grisoli, M. Patrini, G.S. De Magistris, R. Zanoni, Self-assembled monolayers of silver nanoparticles firmly grafted on glass surfaces: Low Ag+ release for an efficient antibacterial activity, J. Colloid. Interface. Sci., 350 (2010) 110-116. https://doi.org/10.1016/j.jcis.2010.06.019

[83] S. Pal, Y.K. Tak, J.M. Song, Does the antibacterial activity of silver nanoparticles depend on the shape of the nanoparticles? a study of the gram-negative bacterium Eschericia coli, Appl. Environ. Microbiol., 73 (2007) 1712-1720. https://doi.org/10.1128/AEM.02218-06

[84] A. Panacek, L. Kvitek, R. Prucek, M. Kolar, R. Vecerova, N. Pizurove, V.K. Sharma, T. Nevecna, R. Zboril, Silver colloid nanoparticles: Synthesis, characterization, and their antibacterial activity, J. Phys. Chem., B. 110 (2006) 16248-16253. https://doi.org/10.1021/jp063826h

[85] G.A. Sotiriou, S.E. Pratsinis, Antibacterial activity of nanosilver ions and particles, Environ. Sci. Technol., 44 (2010) 5649-5654. https://doi.org/10.1021/es101072s

[86] G. Danscher, L. Jansons-Locht, In vivo liberation of silver ions from metallic silver surfaces, Histochem. Cell. Biol., 133 (2010) 359-366. https://doi.org/10.1007/s00418-009-0670-5

[87] C. Ho, S.K. Yau, C. Lok, M. So, C. Che, Oxidative dissolution of silver nanoparticles by biologically relevant oxidants: a kinetic and mechanistic study, Chem. Asian. J., 5 (2010) 285-293. https://doi.org/10.1002/asia.200900387

[88] I. Sondi, B. Salopek-Sondi, Silver nanoparticles as antimicrobial agent: A case on
      E. coli as a model for gram-negative bacteria, J. Colloid. Interface. Sci., 275
      (2004) 177-182. https://doi.org/10.1016/j.jcis.2004.02.012

[89] M. Mayr, M.J. Kim, D. Warner, H. Hopfer, J. Schroeder, M. J. Mihatsch, Argyria
      and decreased kidney function: are silver compounds toxic to the kidney? Am. J.
      Kidney. Dis., 53 (2009) 890-894. https://doi.org/10.1053/j.ajkd.2008.08.028

[90] P.V. Asharani, G.LK. Mun, M.P. Hande, S. Valiyaveettil, Cytotoxicity and
      genotoxicity of silver nanoparticles in human cells, ACS Nano, 3 (2009) 279-290.
      https://doi.org/10.1021/nn800596w

[91] S. Silver, Bacterial silver resistance: Molecular biology and uses and misuses of
      silver compounds, FEMS. Microbiol. Rev., 27 (2003) 341-353.
      https://doi.org/10.1016/S0168-6445(03)00047-0

[92] S.L. Percival, P.G. Bowler, D. Russell, Bacterial resistance to silver in wound
      care. J. Hosp. Infect., 60 (2005) 1-7. https://doi.org/10.1016/j.jhin.2004.11.014

[93] A. Cuin, A.C. Massabni, C.Q.F. Leite, D.N. Sato, A. Neves, B. Szpoganicz, M.S.
      Silve, A.J. Bortoluzzi, Synthesis, X-ray structure and antimycobacterial activity of
      silver complexes with α-hydroxycarboxylic acids, J. Inorg. BioChem., 101 (2007)
      291-296. https://doi.org/10.1016/j.jinorgbio.2006.10.001

[94] P.L. Drake, K.J. Hazelwood, Exposure-related health effects of silver and silver
      compounds: a review, Ann. Occup. Hyg., 49 (2005) 575-585.

[95] E. Ah, W.S. Lee, K.M. Kim, S.Y. Kim, Occupational generalized argyria after
      exposure to aerosolized silver, J. Dermatol., 35 (2008) 759-760.
      https://doi.org/10.1111/j.1346-8138.2008.00562.x

[96] H.B. Kwon, J.H. Lee, S.H. Lee, A.Y. Lee, J.S. Choi, Y.S. Ahn, A case of argyria
      following colloidal silver ingestion, Ann. Dermatol., 21 (2009) 308-310.
      https://doi.org/10.5021/ad.2009.21.3.308

[97] N.S. Tomi, B. Kränke, W. Aberer, A silver man, Lancet., 363 (2004) 532.
      https://doi.org/10.1016/S0140-6736(04)15540-2

[98] S.C. Hau, S.J. Tuft, Presumed corneal argyrosis from occlusive soft contact lenses:
      a case report., Cornea., 28 (2009) 703-705.
      https://doi.org/10.1097/ICO.0b013e31818f9724

[99] X. Wang, H. Chang, R. Francis, H. Olszowy, P. Liu, M. Kempf, L. Cuttle, O.
      Kravchuk, G.E. Phillips, R.M. Kimble, Silver deposits in cutaneous burn scar
      tissue is a common phenomenon following application of a silver dressing, J.

Cutan. Pathol., 36 (2009) 788-792. https://doi.org/10.1111/j.1600-0560.2008.01141.x

[100] V. Karpakam, K. Kamaraj, S. Sathiyanarayanan, G. Venkatachari, S. Ramu, Electrosynthesis of polyaniline–molybdate coating on steel and its corrosion protection performance, Electrochim. Acta, 56 (2011) 2165-2173. https://doi.org/10.1016/j.electacta.2010.11.099

[101] R. Arefinia, A. Shojaei, H. Shariatpanahi, J. Neshati, Anticorrosion properties of smart coating based on polyaniline nanoparticles/epoxy-ester system, Prog. Org. Coat., 75 (2012) 502-508. https://doi.org/10.1016/j.porgcoat.2012.06.003

[102] R. Akid, M, Gobara, H. Wang, Corrosion protection performance of novel hybrid polyaniline/sol–gel coatings on an aluminium 2024 alloy in neutral, alkaline and acidic solutions, Electrochim. Acta, 56 (2011) 2483-2492. https://doi.org/10.1016/j.electacta.2010.12.032

[103] D. Kowalski, M. Ueda, T. Ohtsuka Self-healing ion-permselective conducting polymer coating, J. Mater. Chem., 20 (2010) 7630-7633. https://doi.org/10.1039/c0jm00866d

[104] Z. Zhang, Y. Hu, Z. Liu, T. Guo, Synthesis and evaluation of a moisture-promoted healing copolymer, Polymer, 53 (2012) 2979-2990. https://doi.org/10.1016/j.polymer.2012.04.048

[105] T.F. Da Conceicao, N. Scharnagl, W. Dietzel, D. Hoeche, K.U. Kainer, Study on the interface of PVDF coatings and HF-treated AZ31 magnesium alloy: determination of interfacial interactions and reactions with self-healing properties, Corros. Sci., 53 (2011) 712-719. https://doi.org/10.1016/j.corsci.2010.11.001

[106] B. A¨ıssa, R. Nechache, E. Haddad, W. Jamroz, P.G. Merle, F. Rosei, Ruthenium Grubbs' catalyst nanostructures grown by UV-excimer-laser ablation for self-healing applications, Appl. Surf. Sci., 258 (2012) 9800-9804. https://doi.org/10.1016/j.apsusc.2012.06.032

[107] A. Yabuki, K. Okumura, Self-healing coatings using superabsorbent polymers for corrosion inhibition in carbon steel, Corros. Sci., 59 (2012) 258-262. https://doi.org/10.1016/j.corsci.2012.03.007

[108] A. Yabuki, W. Urushihara, J. Kinugasa, K. Sugano, Self-healing properties of TiO2 particle–polymer composite coatings for protection of aluminum alloys against corrosion in seawater, Mater. Corros., 62 (2011) 907-912. https://doi.org/10.1002/maco.201005756

[109] D.O. Grigoriev, K. Ko"hler, E. Skorb, D.G. Shchukin, H. Mo"hwald, Polyelectrolyte complexes as a "smart" depot for self-healing anticorrosion coatings, Soft. Matter., 5 (2009) 1426-1432. https://doi.org/10.1039/b815147d

[110] G. Williams, S. Geary, H.N. McMurray, Smart release corrosion inhibitor pigments based on organic ion-exchange resins, Corros. Sci., 57 (2012) 139-147. https://doi.org/10.1016/j.corsci.2011.12.024

[111] A. Yabuki, T. Nishisaka, Self-healing capability of porous polymer film with corrosion inhibitor inserted for corrosion protection, Corros. Sci., 53 (2011) 4118-4123. https://doi.org/10.1016/j.corsci.2011.08.022

[112] J. Carneiro, J. Tedim, S.C.M. Fernandes, C.R.S. Freire, A.J.D. Silvestre, A. Gandini, MG.S. Ferreira M.L. Zheludkevich, Chitosan-based self-healing protective coatings doped with cerium nitrate for corrosion protection of aluminum alloy 2024, Prog. Org. Coat., 75 (2012) 8-13. https://doi.org/10.1016/j.porgcoat.2012.02.012

[113] J. Yuan, X. Fang, L. Zhang, G. Hong, Y. Lin, Q. Zheng, Y. Xu, Y. Ruan, W. Weng, H. Xia, G Chen, Multi-responsive self-healing metallo-supramolecular gels based on "click" ligand, J. Mater. Chem., 22 (2012) 11515-11522. https://doi.org/10.1039/c2jm31347b

[114] S. Billiet, W. Van Camp, X.K.D. Hillewaere, H, Rahier, F.E. Du Prez, Development of optimized autonomous self-healing systems for epoxy materials based on maleimide chemistry, Polymer, 53 (2012) 2320-2326. https://doi.org/10.1016/j.polymer.2012.03.061

[115] W. Feng, S.H. Patel, M.Y. Young, J.L. III. Zunino, M. Xanthos, Smart polymeric coatings—recent advances, Adv. Polym. Technol., 26 (2007) 1-13. https://doi.org/10.1002/adv.20083

[116] T. Siva, K. Kamaraj, V. Karpakam, S. Sathiyanarayanan, Soft template synthesis of poly(o-phenylenediamine) nanotubes and its application in self healing coatings, Prog. Org. Coat., 76 (2013) 581-588. https://doi.org/10.1016/j.porgcoat.2012.11.009

[117] M.L. Zheludkevich, J. Tedim, C.S.R. Freire, S.C.M. Fernandes, S. Kallip, A. Lisenkov, A. Gandini, M.G.S. Ferreira, Self-healing protective coatings with "green" chitosan based pre-layer reservoir of corrosion inhibitor, J. Mater. Chem., 21 (2011) 4805-4812. https://doi.org/10.1039/c1jm10304k

[118] D.V. Andreeva, D. Fix, H. Mo¨hwald, D.G. Shchukin, self-healing anticorrosion coatings based on pH-sensitive polyelectrolyte/inhibitor sandwichlike nanostructures, Adv. Mater., 20 (2008) 2789-2794. https://doi.org/10.1002/adma.200800705

[119] D.V. Andreeva, D. Fix, H. Mo¨hwald, D.G. Shchukin, Buffering polyelectrolyte multilayers for active corrosion protection, J. Mater. Chem., 18 (2008)1738-1740. https://doi.org/10.1039/b801314d

[120] S.J. Garc'ıa, H.R. Fischer, P. White, J. Mardel, Y. Gonza'lez-Garcia, J.M.C. Mol, A.E. Hughes, Self-healing anticorrosive organic coating based on an encapsulated water reactive silyl ester: Synthesis and proof of concept, Prog. Org. Coat., 70 (2011) 142-149. https://doi.org/10.1016/j.porgcoat.2010.11.021

[121] G. Williams, A. Gabriel, A. Cook, H.N. McMurray, Dopant Effects in Polyaniline Inhibition of Corrosion-Driven Organic Coating Cathodic Delamination on Iron, J. Electrochem. Soc., 153 (2006) B425-B433. https://doi.org/10.1149/1.2229280

[122] T. Tu¨ken, B. Yazıcı, M. Erbil, Dopant Effects in Polyaniline Inhibition of Corrosion-Driven Organic Coating Cathodic Delamination on Iron, Appl. Surf. Sci., 252 (2006) 2311-2318.

[123] D. Borisova, H. Mo¨hwald, D.G. Shchukin, Influence of Embedded Nanocontainers on the Efficiency of Active Anticorrosive Coatings for Aluminum Alloys Part II: Influence of Nanocontainer Position, ACS Appl. Mater. & Interf., 5 (2013) 80-87. https://doi.org/10.1021/am302141y

[124] I.A. Kartsonakis, A.C. Balaskas, E.P. Koumoulos, C.A. Charitidis, G. Kordas, Evaluation of corrosion resistance of magnesium alloy ZK10 coated with hybrid organic–inorganic film including containers, Corros. Sci., 65 (2012) 481-493. https://doi.org/10.1016/j.corsci.2012.08.052

[125] A. Stankiewicz, M. Barker, Development of self-healing coatings for corrosion protection on metallic structures, Smart Mater. Struct., 25 (2016) 1-10. https://doi.org/10.1088/0964-1726/25/8/084013. https://doi.org/10.1088/0964-1726/25/8/084013

[126] M.L. Zheludkevich, J. Tedim, M.G.S. Ferreira, "Smart" coatings for active corrosion protection based on multi-functional micro and nanocontainers, Electrochim. Acta, 82 (2012) 314-323. https://doi.org/10.1016/j.electacta.2012.04.095

[127] D,G, Shchukin, M.L. Zheludkevich, H. Mo¨hwald Feedback active coatings based on incorporated nanocontainers, J. Mater. Chem., 16 (2006) 4561-4566. https://doi.org/10.1039/B612547F

[128] T.C. Mauldin, M.R. Kessler, Self-healing polymers and composites. Int. Mater. Rev., 55 (2010) 317-346. https://doi.org/10.1179/095066010X12646898728408

[129] T. Nesterova, K. Dam-Johansen, L.T. Pedersen, S. Kiil, Microcapsule-based self-healing anticorrosive coatings: Capsule size, coating formulation, and exposure testing, Prog. Org. Coat., 75 (2012) 309-318. https://doi.org/10.1016/j.porgcoat.2012.08.002

[130] N. Selvakumar, K. Jeyasubramanian, R. Sharmila, Smart coating for corrosion protection by adopting nano particles, Prog. Org. Coat., 74 (2012) 461-469. https://doi.org/10.1016/j.porgcoat.2012.01.011

[131] X. Liu, H. Zhang, J. Wang, Z. Wang, S. Wang, Preparation of epoxy microcapsule based self-healing coatings and their behaviour, Surf. Coat. Technol., 206 (2012) 4976-4980. https://doi.org/10.1016/j.surfcoat.2012.05.133

[132] T. Nesterova, K. Dam-Johansen, S. Kiil, Synthesis of durable microcapsules for self-healing anticorrosive coatings: A comparison of selected methods, Prog. Org. Coat., 70 (2011) 342-352. https://doi.org/10.1016/j.porgcoat.2010.09.032

[133] R.S. Jadhav, D.G. Hundiwale, P.P. Mahulikar, Synthesis and characterization of phenol–formaldehyde microcapsules containing linseed oil and its use in epoxy for self-healing and anticorrosive coating, J. Appl. Polym. Sci., 119 (2011) 2911-2916. https://doi.org/10.1002/app.33010

[134] Z. Yang, Z. Wei, L. Liao, H. Wang, W. Li, The self-healing composite anticorrosion coating, Physics. Procedia., 18 (2011) 216-221. https://doi.org/10.1016/j.phpro.2011.06.084

[135] Y. Zhao, W. Zhang, L. Liao, S. Wang, W. Li, Self-healing coatings containing microcapsule, Appl. Surf. Sci., 258 (2012) 1915-1918. https://doi.org/10.1016/j.apsusc.2011.06.154

[136] M. Huang, J. Yang, Facile microencapsulation of HDI for self-healing anticorrosion coatings, J. Mater. Chem., 21 (2011) 11123-11130. https://doi.org/10.1039/c1jm10794a

[137] H. Choi, Y.K. Song, K.Y. Kim, J.M. Park, Encapsulation of triethanolamine as organic corrosion inhibitor into nanoparticles and its active corrosion protection

for steel sheets, Surf. Coat. Technol., 206 (2012) 2354-2362.
https://doi.org/10.1016/j.surfcoat.2011.10.030

[138] D.G. Shchukin, D.O. Grigoriev, H. Mo¨hwald, Application of smart organic nanocontainers in feedback active coatings, Soft. Matter., 6 (2010) 720-725. https://doi.org/10.1039/B918437F

[139] D.G. Shchukin, H. Mo¨hwald, Smart nanocontainers as depot media for feedback active coatings, Chem. Commun., 47 (2011) 8730-8739. https://doi.org/10.1039/c1cc13142g

[140] A. Pilba'th, T. Szabo´, J. Telegdi, L. Nyikos, SECM study of steel corrosion under scratched microencapsulated epoxy resin, Prog. Org. Coat., 75 (2012) 480-485. https://doi.org/10.1016/j.porgcoat.2012.06.006

[141] M.F. Montemor, D.V. Snihirova, M.G. Taryba, S.V. Lamaka, I.A. Kart-sonakis, A.C. Balaskas, G.C. Kordas, J. Tedim, A. Kuznetsova, M.L. Zheludkevich, M.G.S. Ferreira Evaluation of self-healing ability in protective coatings modified with combinations of layered double hydroxides and cerium molibdate nanocontainers filled with corrosion inhibitors, Electrochim. Acta, 60 (2012) 31-40. https://doi.org/10.1016/j.electacta.2011.10.078

[142] J. Tedim, S.K. Poznyak, A. Kuznetsova, D. Raps, T. Hack, M.L. Zheludkevich, M.G.S. Ferreira, Enhancement of active corrosion protection via combination of inhibitor-loaded nanocontainers, ACS Appl. Mater. Interfac., 2 (2010) 1528-1535. https://doi.org/10.1021/am100174t

[143] M.L. Zheludkevich, S.K. Poznyak, L.M. Rodrigues, D. Raps, T. Hackc, L.F. Dick, T. Nunes, M.G.S. Ferreira, Active protection coatings with layered double hydroxide nanocontainers of corrosion inhibitor, Corros. Sci., 52 (2010) 602-611. https://doi.org/10.1016/j.corsci.2009.10.020

[144] M.F. Haase, D.O. Grigoriev, H. Mo¨hwald, D.G. Shchukin, development of nanoparticle stabilized polymer nanocontainers with high content of the encapsulated active agent and their application in water-borne anticorrosive coatings, Adv. Mater., 24 (2012) 2429-2435. https://doi.org/10.1002/adma.201104687

[145] F. Maia, J. Tedim, A.D. Lisenkov, A.N. Salak, M.L. Zheludkevich, M.G.S. Ferreira, Silica nanocontainers for active corrosion protection, Nanoscale, 4 (2012) 1287-1298. https://doi.org/10.1039/c2nr11536k

[146] D.G. Shchukin, M. Zheludkevich, K. Yasakau, S. Lamaka, M.G.S. Ferreira, H. Mo¨hwald, Layer-by-layer assembled nanocontainers for self-healing corrosion protection, Adv. Mater., 18 (2006) 1672-1678. https://doi.org/10.1002/adma.200502053

[147] H. Gro¨ger, F. Gyger, P. Leidinger, C. Zurmu¨hl, C. Feldmann, Microemulsion approach to nanocontainers and its variability in composition and filling, Adv. Mater., 21 (2009)1586-1590. https://doi.org/10.1002/adma.200802972

[148] S.V. Lamaka, M.L. Zheludkevich, K.A. Yasakau, F.M. Montemor, P. Cecilio, M.G.S. Ferreira, TiOx self-assembled networks prepared by templating approach as nanostructured reservoirs for self-healing anticorrosion pre-treatments, Electrochem. Commun., 8 (2006) 421-428. https://doi.org/10.1016/j.elecom.2005.12.019

[149] A.C. Balaskas, I.A. Kartsonakis, L.A. Tziveleka, G.C. Kordas, Improvement of anti-corrosive properties of epoxy-coated AA 2024-T3 with TiO_2 nanocontainers loaded with 8-hydroxyquinoline, Prog. Org. Coat., 74 (2012) 418-426. https://doi.org/10.1016/j.porgcoat.2012.01.005

[150] D. Snihirova, S.V. Lamaka, M.F. Montemor, "SMART" protective ability of water based epoxy coatings loaded with $CaCO_3$ microbeads impregnated with corrosion inhibitors applied on AA2024 substrates, Electrochim. Acta, 83 (2012) 439-447. https://doi.org/10.1016/j.electacta.2012.07.102

[151] M. Kouhi, A. Mohebbi, M. Mirzaei, Evaluation of the corrosion inhibition effect of micro/nanocapsulated polymeric coatings: a comparative study by use of EIS and Tafel experiments and the area under the Bode plot, Res. Chem. InterMed., 39 (2012) 2049-2062. https://doi.org/10.1007/s11164-012-0736-1

[152] G.L. Li, Z. Zheng, H. Mo¨hwald, D.G. Shchukin, Silica/Polymer Double-Walled Hybrid Nanotubes: Synthesis and Application as Stimuli-Responsive Nanocontainers in Self-Healing Coatings, ACS Nano, 7 (2013) 2470-2478. https://doi.org/10.1021/nn305814q

[153] D. Fix, D.V. Andreeva, Y.M. Lvov, D.G. Shchukin, H. Mo¨hwald, Application of inhibitor-loaded halloysite nanotubes in active anti-corrosive coatings, Adv. Funct. Mater., 19 (2009) 1720-1727. https://doi.org/10.1002/adfm.200800946

[154] N.P. Tavandashti, S. Sanjabi, Corrosion study of hybrid sol–gel coatings containing boehmite nanoparticles loaded with cerium nitrate corrosion inhibitor, Prog. Org. Coat., 69 (2010) 384-391. https://doi.org/10.1016/j.porgcoat.2010.07.012

[155] Y.M. Lvov, D.G. Shchukin, H. Mo¨hwald, R.R. Price, Halloysite clay nanotubes for controlled release of protective agents, J. Nanosci. Nanotech., 2 (2008) 814-820. https://doi.org/10.1021/nn800259q

[156] X. Jiang, Y.B. Jiang, N. Liu, H. Xu, S. Rathod, P. Shah, C.F. Binker, Controlled release from core-shell nanoporous silica particles for corrosion inhibition of aluminum alloys, J. Nanomater., 2011 (2011) 1-10. https://doi.org/10.1155/2011/760237

[157] M. Zheludkevich, R. Serra, M. Montemor, M. Ferreira, Oxide nanoparticle reservoirs for storage and prolonged release of the corrosion inhibitors, Electrochem. Commun., 7 (2005) 836-840. https://doi.org/10.1016/j.elecom.2005.04.039

[158] S.V. Lamaka, M.L. Zheludkevich, K.A. Yasakau, R. Serra, S.K. Poznyak, M.G.S. Ferreira, Nanoporous titania interlayer as reservoir of corrosion inhibitors for coatings with self-healing ability, Prog. Org. Coat., 58 (2007) 127-135. https://doi.org/10.1016/j.porgcoat.2006.08.029

[159] A.N. Khramov, N.N. Voevodin, V.N. Balbyshev, M.S. Donley, Hybrid organo-ceramic corrosion protection coatings with encapsulated organic corrosion inhibitors, Thin. Solid. Films, 447–448 (2004) 549-557. https://doi.org/10.1016/j.tsf.2003.07.016

[160] D. Borisova, H. Mo¨hwald, D.G. Shchukin, Influence of embedded nanocontainers on the efficiency of active anticorrosive coatings for aluminum alloys part I: influence of nanocontainer concentration, Appl. Mater. Interf., 4 (2012) 2931-2939. https://doi.org/10.1021/am300266t

[161] A. Latnikova, D.O. Grigoriev, J. Hartmann, H. Mo¨hwald, D.G. Shchukin, Polyfunctional active coatings with damage-triggered water-repelling effect, Soft. Matter., 7 (2011) 369-372. https://doi.org/10.1039/C0SM00842G

[162] C. Suryanarayana, K. C. Rao, D. Kumar, Preparation and characterization of microcapsules containing linseed oil and its use in self-healing coatings, Prog. Org. Coat., 63 (2008) 72-78. https://doi.org/10.1016/j.porgcoat.2008.04.008

[163] J. Fickert, C. Wohnhaas, A. Turshatov, K. Landfester, D. Crespy, Copolymers structures tailored for the preparation of nanocapsules, Macromolecul., 46 (2013) 573-597. https://doi.org/10.1021/ma302013s

[164] V. Sauvant-Moynot, S. Gonzalez, J. Kittel, Self-healing coatings: An alternative route for anticorrosion protection, Prog. Org. Coat., 63 (2008) 307-315. https://doi.org/10.1016/j.porgcoat.2008.03.004

[165] A. Aramaki, Synergistic inhibition of zinc corrosion in 0.5 M NaCl by combination of cerium(III) chloride and sodium silicate, Corros. Sci., 44 (2002) 871-886. https://doi.org/10.1016/S0010-938X(01)00087-7

[166] M.L. Zheludkevich, D.G. Shchukin, K.A. Yasakau, H. Mo"hwald, M.G.S. Ferreira, Anticorrosion coatings with self-healing effect based on nanocontainers impregnated with corrosion inhibitor, Chem. Matter., 19 (2007) 402-411. https://doi.org/10.1021/cm062066k

[167] P.D. Tatiya, R.K. Hedaoo, P.P. Mahulikar, V.V. Gite, Novel polyurea microcapsules using dendritic functional monomer: synthesis, characterization, and its use in self-healing and anticorrosive polyurethane coatings, Ind. Eng. Chem. Res., 52 (2013) 1562-1570. https://doi.org/10.1021/ie301813a

[168] D. Borisova, H. Mo"hwald, D.G. Shchukin, Mesoporous silica nanoparticles for active corrosion protection, ACS Nano, 5 (2011) 1939-1946. https://doi.org/10.1021/nn102871v

[169] E.M. Moustafa, A. Dietz, T. Hochsattel, Manufacturing of nickel/nanocontainer composite coatings, Surf. Coat. Technol., 216 (2013) 93-99. https://doi.org/10.1016/j.surfcoat.2012.11.030

[170] M. Samadzadeh, S.H. Boura, M. Peikari, S. M. Kasiriha, A. Ashrafi, A review on self-healing coatings based on micro/nanocapsules, Prog. Org. Coat., 68 (2010) 159-164. https://doi.org/10.1016/j.porgcoat.2010.01.006

[171] A. Kumar, L.D. Stephenson, J.N. Murray, Self-healing coatings for steel, Prog. Org. Coat., 55 (2006) 244-253. https://doi.org/10.1016/j.porgcoat.2005.11.010

[172] A. Latnikova, D. Grigoriev, M. Schenderlein, H. Mo"hwald, D. Shchukin, A new approach towards "active" self-healing coatings: exploitation of microgels, Soft. Matter., 8 (2012) 10837-10844. https://doi.org/10.1039/c2sm26100f

[173] N.C. Rosero-Navarro, L. Paussa, F. Andreatta, Y. Castro, A. Duran, M. Aparicio, L. Fedrizzi, Optimization of hybrid sol–gel coatings by combination of layers with complementary properties for corrosion protection of AA2024, Prog. Org. Coat., 69 (2010) 167-174. https://doi.org/10.1016/j.porgcoat.2010.04.013

[174] N.C. Rosero-Navarro, S.A. Pellice, A. Dura'n, M. Aparicio, Effects of Ce-containing sol–gel coatings reinforced with SiO2 nanoparticles on the protection

of AA2024. Corros. Sci., 50 (2008) 1283-1291.
https://doi.org/10.1016/j.corsci.2008.01.031

[175] F. Andreatta, P. Aldighieri, L. Paussa, R. Di Maggio, S. Rossi, L. Fedrizzi, Electrochemical behaviour of ZrO2 sol–gel pre-treatments on AA6060 aluminium alloy, Electrochim. Acta, 52 (2007) 7545–7555. https://doi.org/10.1016/j.electacta.2006.12.065

[176] M.F. Montemor, W. Trabelsi, S.V. Lamaka, K.A. Yasakau, M.L. Zheludkevich, A.C. Bastos, M.G.S. Ferreira, The synergistic combination of bis-silane and CeO2·ZrO2 nanoparticles on the electrochemical behaviour of galvanised steel in NaCl solutions, Electrochim. Acta, 53 (2008) 5913-5922. https://doi.org/10.1016/j.electacta.2008.03.069

[177] M.F. Montemor, R. Pinto, M.G.S. Ferreira, Chemical composition and corrosion protection of silane films modified with CeO2 nanoparticles, Electrochim. Acta, 54 (2009) 5179-5189. https://doi.org/10.1016/j.electacta.2009.01.053

[178] L.K. Wu, L. Liu, J. Li, J.M. Hu, J.Q. Zhang, C.N. Cao, Electrodeposition of cerium (III)-modified bis-[triethoxysilypropyl]tetra-sulphide films on AA2024-T3 (aluminum alloy) for corrosion protection, Surf. Coat. Technol., 204 (2010) 3920-3926. https://doi.org/10.1016/j.surfcoat.2010.05.027

[179] J. Bardon, J. Bour, H. Aubriet, D. Ruch, B. Verheyde, R. Dams, S. Paulussen, R. Rego, D. Vangeneugden, Deposition of Organosilicon-based anticorrosion layers on galvanized steel by atmospheric pressure dielectric barrier discharge plasma. Plasma Proc. Polym., 4 (2007) S445-S449. https://doi.org/10.1002/ppap.200731204

[180] L. Paussa, N.C. Rosero-Navarro, F. Andreatta, Y. Castro, A. Duran, M. Aparicio, L. Fedrizzi, Inhibition effect of cerium in hybrid sol–gel films on aluminium alloy AA2024, Surf. Interface. Anal., 42 (2010) 299-305. https://doi.org/10.1002/sia.3198

[181] A. Pepe, M. Aparicio, S. Cere´, A. Dura'n, Preparation and characterization of cerium doped silica sol–gel coatings on glass and aluminum substrates J. Non-Cryst. Solids, 348 (2004) 162-171. https://doi.org/10.1016/j.jnoncrysol.2004.08.141

[182] A. Pepe, M. Aparicio, A. Dura'n, S. Cere´, Cerium hybrid silica coatings on stainless steel AISI 304 substrate, J. Sol-Gel. Sci. Technol., 39 (2006) 131-138. https://doi.org/10.1007/s10971-006-9173-1

[183] F. Andreatta, L. Paussa, P. Aldighieri, A. Lanzutti, D. Raps, L. Fedrizzi, Corrosion behaviour of sol–gel treated and painted AA2024 aluminium alloy, Prog. Org. Coat., 69 (2010) 133-142. https://doi.org/10.1016/j.porgcoat.2010.04.012

[184] L. Paussa, N.C. Rosero Navarro, D. Bravin, F. Andreatta, A. Lanzutti, M. Aparicio, A. Duran, L. Fedrizzi, ZrO2 sol–gel pre-treatments doped with cerium nitrate for the corrosion protection of AA6060, Prog. Org. Coat., 74 (2012) 311-319. https://doi.org/10.1016/j.porgcoat.2011.08.017

[185] N.C. Rosero-Navarro, M. Curioni, Y. Castro, M. Aparicio, G.E. Thompson, A. Duran, Glass-like CexOy sol–gel coatings for corrosion protection of aluminium and magnesium alloys, Surf. Coat. Technol., 206 (2011) 257-264. https://doi.org/10.1016/j.surfcoat.2011.07.006

[186] N.C. Rosero-Navarro, S.A. Pellice, A. Dura'n, S. Cere', M. Aparicio, Corrosion protection of aluminium alloy AA2024 with cerium doped methacrylate-silica coatings, J. Sol-Gel. Sci. Technol., 52 (2009) 31-40. https://doi.org/10.1007/s10971-009-2010-6

[187] C. Motte, M. Poelman, A. Roobroeck, M. Fedel, F. Deflorian, M.G. Olivier, Improvement of corrosion protection offered to galvanized steel by incorporation of lanthanide modified nanoclays in silane layer, Prog. Org. Coat., 74 (2012) 326-333. https://doi.org/10.1016/j.porgcoat.2011.12.001

[188] A. Collazo, A. Covelo, X.R. No'voa, C. Pe'rez, Corrosion protection performance of sol–gel coatings doped with red mud applied on AA2024-T3, Prog. Org. Coat., 74 (2012) 334-342. https://doi.org/10.1016/j.porgcoat.2011.10.001

[189] S.M. Madani, M. Ehteshamzadeh, H.H. Rafsanjani, S.S. Mansoori, The effect of calcination on the corrosion performance of TiO2 sol–gel coatings doped with benzotriazole on steel CK45, Mater. Corros. 61(2009) 318-323.

[190] S.M. Hosseini, A.H. Jafari, E. Jamalizadeh, Self-healing corrosion protection by nanostructure sol–gel impregnated with propargyl alcohol Electrochim. Acta, 54 (2009) 7207-7213. https://doi.org/10.1016/j.electacta.2009.07.002

[191] W. Trabelsi, P. Cecilio, M.G.S Ferreira, M.F. Montemor, Electrochemical assessment of the self-healing properties of Ce-doped silane solutions for the pre-treatment of galvanised steel substrates, Prog. Org. Coat., 54 (2005) 276-284. https://doi.org/10.1016/j.porgcoat.2005.07.006

[192] K.A. Yasakau, M.L. Zheludkevich, O.V. Karavai, M.G.S Ferreira, Influence of inhibitor addition on the corrosion protection performance of sol–gel coatings on

AA2024, Prog. Org. Coat., 63 (2008) 352-361.
https://doi.org/10.1016/j.porgcoat.2007.12.002

[193] M. Zaharescua, L.A. Predoana, A. Barau, D. Raps, F. Gammel, N.C. Rosero-Navarro, Y. Castro, A. Duran, M. Aparicio, SiO2 based hybrid inorganic–organic films doped with TiO2–CeO2 nanoparticles for corrosion protection of AA2024 and Mg-AZ31B alloys, Corros. Sci., 51 (2009) 1998-2005.
https://doi.org/10.1016/j.corsci.2009.05.022

[194] E. Roussi, A. Tsetsekou, D. Tsiourvas, A. Karantonis, Novel hybrid organo-silicate corrosion resistant coatings based on hyperbranched polymers, Surf. Coat. Technol., 205 (2011) 3235-3244. https://doi.org/10.1016/j.surfcoat.2010.11.037

[195] L.M. Palomino, P.H. Suegama, I.V. Aoki, F.M. Montemor, H.G. De Melo, Electrochemical study of modified cerium–silane bi-layer on Al alloy 2024-T3, Corros. Sci., 51 (2009) 1238-1250. https://doi.org/10.1016/j.corsci.2009.03.012

[196] M.F. Montemor, M.G.S. Ferreira, Analytical characterization of silane films modified with cerium activated nanoparticles and its relation with the corrosion protection of galvanised steel substrates, Prog. Org. Coat., 63 (2008) 330-337.
https://doi.org/10.1016/j.porgcoat.2007.11.008

[197] S. Li, J. Fu, Improvement in corrosion protection properties of TiO2 coatings by chromium doping, Corros. Sci., 68 (2013) 101-110.
https://doi.org/10.1016/j.corsci.2012.10.040

[198] D. Zhu, W.J. van Ooij, Corrosion protection of AA 2024-T3 by bis-[3-(triethoxysilyl)propyl] tetrasulfide in sodium chloride solution: Part 2: mechanism for corrosion protection, Corros. Sci., 45 (2003) 2177-2197.
https://doi.org/10.1016/S0010-938X(03)00061-1

[199] A.S. Hamdy, I. Doench, H. Mo"hwald, Intelligent self-healing corrosion resistant vanadia coating for AA2024, Thin. Solid. Films, 520 (2011) 1668-1678.
https://doi.org/10.1016/j.tsf.2011.05.080

[200] A.S. Hamdy, I. Doench, H. Mo"hwald, Vanadia-based coatings of self-repairing functionality for advanced magnesium Elektron ZE41 Mg–Zn–rare earth alloy, Surf. Coat. Technol., 206 (2012) 3686-3692.
https://doi.org/10.1016/j.surfcoat.2012.03.025

[201] M. Yuan, J. Lu, G. Kong, C. Che, Self healing ability of silicate conversion coatings on hot dip galvanized steels, Surf. Coat. Technol., 205 (2011) 4507-4513.
https://doi.org/10.1016/j.surfcoat.2011.03.088

[202] A.S. Hamdy, I. Doench, H. Mo"hwald, Assessment of a one-step intelligent self-healing vanadia protective coatings for magnesium alloys in corrosive media, Electrochim. Acta, 56 (2011) 2493-2502. https://doi.org/10.1016/j.electacta.2010.11.103

[203] A.S. Hamdy, D.P. Butt, Novel smart stannate based coatings of self-healing functionality for AZ91D magnesium alloy, Electrochim. Acta, 97 (2013) 296-303. https://doi.org/10.1016/j.electacta.2013.02.108

[204] A. Yabuki, R.Kaneda, Barrier and self-healing coating with fluoro-organic compound for zinc. Mater. Corros. 60 (2009) 444-449. https://doi.org/10.1002/maco.200805100

[205] A. Yabuki, M. Sakai, Self-healing coatings of inorganic particles using a pH-sensitive organic agent, Corros. Sci., 53 (2011) 829-833. https://doi.org/10.1016/j.corsci.2010.11.021

[206] H.J. Yang, Y.T. Pei, J.C. Rao, J.T.M. De Hosson, Self-healing performance of Ti2AlC ceramic, J. Mater. Chem., 22 (2012) 8304-8313. https://doi.org/10.1039/c2jm16123k

[207] J. Gao, J. Suo, Effects of heating temperature and duration on the microstructure and properties of the self-healing coatings, Surf. Coat. Technol., 206 (2011) 1342-1350. https://doi.org/10.1016/j.surfcoat.2011.08.059

[208] C.-Y. Kuo, Y.-Y. Chen, S.-Y. Lu, A facile route to create surface porous polymer films via phase separation for antireflection applications, ACS Appl. Mater. Interf. 1 (2008) 72–75. https://doi.org/10.1021/am800002x

[209] H. Shimomura, Z. Gemici, R.E. Cohen, M.F. Rubner, Layer-by-Layer-Assembled High-Performance Broadband Antireflection Coatings, ACS Appl. Mater. Interf. 2 (2010) 813–820. https://doi.org/10.1021/am900883f

[210] F. Cebeci, Z. Wu, L. Zhai, R.E. Cohen, M.F. Rubner, Nanoporosity-driven superhydrophilicity: a means to create multifunctional antifogging coatings, Langmuir., 22 (2006) 2856–2862. https://doi.org/10.1021/la053182p

[211] J.A. Howarter, J.P. Yougblood, self-cleaning and next generation anti-fog surfaces and coatings, Macromol. Rapid Commun., 29 (2008) 455–466. https://doi.org/10.1002/marc.200700733

[212] L. Zhang, Y. Li, J. Sun, J. Shen, Mechanically stable antireflection and antifogging coatings fabricated by the layer-by-layer deposition process and postcalcination, Langmuir., 24 (2008) 10851–10857. https://doi.org/10.1021/la801806r

[213] S. Grube, K. Siegmann, M. Hirayama, A moisture-absorbing and abrasion-resistant transparent coating on polystyrene, J. Coat. Technol. Res., 12 (2015) 669-680. https://doi.org/10.1007/s11998-015-9678-z

[214] W.S. Creasy, Hydrophilic polyvinylbutyral alloys. U.S. Patent 4, 847, 324, Juy 11, 1989.

[215] M. Funaki, M. Yoshida, Y. Shimauchi, A. Fujioka, K. Sakiyama, Coated materials and production thereof. U.S. Patent 4, 242, 412, December 30, 1980.

[216] I.J. Haller, Covalently attached organic monolayers on semiconductor surfaces, Am. Chem. Soc., 100 (1978) 8050–8055. https://doi.org/10.1021/ja00494a003

[217] H. Hosono, T. Taniguchi, Anti-Fogging Film, U.S. Patent 5, 134, 021, July 28, 1992.

[218] H. Hosono, T. Taniguchi, M. Nishii, Process for Preparation of Anti-Fogging Coating, U.S. Patent 5, 075, 133, December 24, 1991.

[219] B.L. Laurin, Abrasion and antifog-resistant optical element, U.S. Patent 4, 127, 682, November 28, 1978.

[220] N.E. Petersen, Anti-fogging surgical mask, U.S. Patent 4, 419, 993, December 13, 1983.

[221] J.A.J. Sanders, M.J. Larson, Transparent anti-fog compositions, U.S. Patent 4,615,738, October 7, 1986.

[222] M. Haga, Y. Onisawa, K. Shimizu, Plastic lenses and method of producing the same, U.S. Patent 5, 985, 420, November 16, 1999.

[223] J.C. Song, Transparent anti-fog coating, U.S. Patent 5, 804, 612, September 8, 1998.

[224] H. Eggers, R. Klein, C. Muller, R. Brandt, Multilayer film with lamination and heat-sealable sides, and having antifogging properties, U.S. Patent 6, 576, 348, December 12, 2002.

[225] H. Lee, M. L. Alcaraz, M. F. Rubner, R. E. Cohen, Zwitter-wettability and antifogging coatings with frost-resisting capabilities, ACS Nano, 7 (2013) 2172–2185. https://doi.org/10.1021/nn3057966

[226] H. Nagasawa, M. Nakamura, M. Ishida, H. Ohsawa, T. Hoga, Antifogging agent composition. U.S. Patent 2003/0127625 A1, July 10, 2003.

[227] S. Yamazaki, N. Murata, H. Yamamoto, Article with antifogging film and process for producing same, U.S. Patent 6,420,020 B1, July 16, 2002.

[228] S. Yamazaki, N. Murata, H. Yamamoto, Article with antifogging film and process for producing same, U.S. Patent 6, 531, 215, March 11, 2003.

[229] A.A. Kruger, P. Chartier. Anti-fogging coating composition, product coated with said composition for preparation of said product, U.S. Patent 5, 578, 378, November 26, 1996.

[230] M. Keller, M. Lenhard, Substrate and polymerizable mixture, method of manufacturing of said polymerizable mixture, and method of manufacturing of a nonfogging or low fogging layer, U.S. Patent 5, 648, 441, July 15, 1997.

[231] Y. Oshibe, Y. Yamamoto, H. Ohmura, K. Kumazawa, Composition of ultraviolet curing antifogging agent and process for forming antifogging coating film, U.S. Patent 5, 244, 935, September 14, 1993.

[232] Y. Oshibe, Y. Yammamoto, H. Ohmura, K. Kumazawa, Anti-fogging resin film-forming composition, U.S. Patent 5, 180, 760, January, 19, 1993.

[233] P. Chevallier, S. Turgeon, C. Sarra-Bournet, R. Turcotte, G. Laroche, Characterization of multilayer anti-fog coatings, ACS Appl. Mater. Interf., 3 (2011) 750–758. https://doi.org/10.1021/am1010964

Chapter 4

Efficient Polymer Decorated Bimetallic Nanosorbents for Dye Removal Applications

Aysun Savk, Betul Sen, Fatih Sen*

Sen Research Group, Biochemistry Department, Faculty of Arts and Science, Dumlupınar University, Evliya Çelebi Campus, 43100 Kütahya, Turkey.

*fatih.sen@dpu.edu.tr

Abstract

Herein, the microwave assisted methodology was employed to produce polyaniline decorated platinum-nickel nanoparticles (Pt-Ni@PANI NPs) as nanosorbents for the methylene blue removal at room temperature via adsorption. These nano sorbents were characterized by XRD, TEM, HRTEM and XPS. The prepared Pt-Ni@PANI NPs displayed highly crystalline, monodisperse and colloidal stable structures. Adsorption measurements represented that one of the highest MB adsorption was obtained as 271.15 mg/g in 55 min, which can be considered a very good capacity. Furthermore, Pt-Ni@PANI NPs are reusable materials for the MB removal application because they sustained 55.6 % of the initial efficiency after six successive adsorptions–desorption cycles.

Keywords

Adsorption, Bimetallic Nanosorbents, Dye Removal, Outstanding Capacity, Polymer Decoration

Contents

1.   Introduction

In numerous industries, dyes are commonly being used; for example, coating, cosmetic, leather, paint, paper, plastic, textile industries, and they have big amounts of effluent water and this effluent water typically contains organic dyes. They are very toxic, and on the other hand, the dyes are visible and, therefore, unaesthetic [1-4]. Therefore, conventional dye removal methods are being used to clean these contaminated effluents [5]. Today, in addition to conventional treatment methods, various techniques such as biological treatment, electrochemical degradation, flocculation, oxidation, photocatalytic degradation, sonochemical degradation, ultrafiltration, and adsorption processes are also being utilized for dye removal from wastewaters, [3, 5-6]. If those techniques are compared, one can see that the adsorption process is the most applicable method for dye removal as it is very efficient, practical, and economic [2, 4, 7]. As adsorption process is a very demandable method for effluent treatments, numerous approaches have been executed to develop various efficient adsorbents. As a result, people have proven that activated carbon, clay materials, chitin, chitosan, peat, silica, and solid waste are good candidates for adsorbents [8, 9]. However they have some problems such as lack of specificity, high costs for their synthesis and treatment, low efficacy and capacity, problems with their recycling and reusability, and longer processing times [2, 4, 9]. Hence, much better adsorbent materials are required for more effective dye removal. Generally, nanomaterials have lately been used as efficacious and reusable materials for many applications [11-32]. They have mechanical flexibility, chemical stability, regulatable pore size, modifiable structures and compositions. Thus, for the remediation of dyes from contaminated waters, different nanomaterials have been produced, modified and improved by researchers. Latterly, carbon supported nanoparticles or carbon nanotubes, polyaniline nanotubes, iron oxide nanoparticles, fullerenes, polyurethane foams, polypyrrole/TiO2 nanocomposites, PZS or poly(cyclotri-phosphazene-co-4,4-sulfonyl-diphenol) nanospheres, polydopamine, nanospheres and CdS nanostructures have been developed [1-10,33]. These materials and their modified kinds are starting to be used in dye removal applications [2,9,10,33]. These nanomaterials have a wide surface area which allows better contact, and hence, better dye removal. All these create ideal characteristics for demandable nanosorbents [10,33-38]. Recently, a conducting polymer, polyaniline (PANI), was tested in adsorption of dye effluent [33-34]. Polyaniline is considered to be one of the most promising classes of organic conducting polymers due to their well-defined electrochemistry, easy protonation reversibility, excellent redox recyclability, and good environmental stability, and variety of nanostructured morphologies [34-38]. It was reported to be utilized as an adsorbent for adsorption of protein and DNA [39-40]. For successful dye removal in effluent waters, better performing nanocomposites are required and research should have a simple adsorbent preparation methodology in addition to the composition of adsorbent [41]. In addition, it is known that homogeneous distribution of nanomaterials on the surface of support materials is a prerequisite to obtain high performance of adsorbents [41]. For this reason, Pt-Ni@PANI NPs were produced through the microwave-assisted method for the first time for the determination of the adsorption capacity towards MB removal. In Scheme 1, the general mechanism of Pt-Ni@PANI NPs is given for MB removal. The relationship between contact time and adsorption efficiency and the relationship between the amount of nanomaterials adsorbed per unit weight of methylene blue and dye concentration, as well as the reusability of nanomaterials were investigated. X-ray diffraction (XRD), transmission electron microscopy (TEM), high resolution transmission electron microscopy (HRTEM) and X-ray photoelectron spectroscopy (XPS) were used to characterize the synthesized Pt-Ni@PANI NPs. The MB removal efficiency was examined via UV-Vis spectrophotometer. It was displayed that the Pt-Ni@PANI NPs is a new type of nanosorbent and it has great properties such as high surface area, fast extraction and regeneration periods, ease of operation, and high potential for remediating wastewater for dye removal applications.

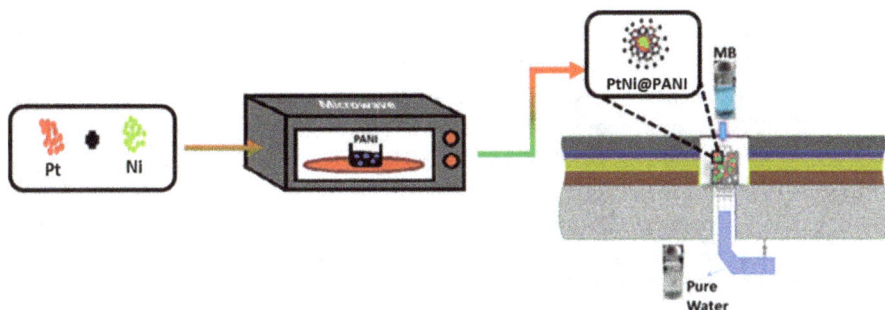

Scheme-1 The mechanism of methylene blue removal with Pt-Ni@PANI NPs. Here, the bottle having blue liquid at the top represents the water polluted with MB, the red and gray circles stand Pt-Ni@PANI NPs and the blue drawing at the bottom is for clean water.

2.   Materials and methods

Aldrich supplied PtCl$_4$, NiCl$_2$, EG and PANI. C$_2$H$_5$OH and water used during this study were provided from Merck and Milli Q-pure machine, respectively. Before washing all glass pieces and other lab materials with a large amount of distilled water, they were cleaned with acetone, and then dried.

2.1   Instrumentation

TEM images of prepared Pt-Ni@PANI NPs have been obtained by a JEOL 200 kV TEM instrument. Sample preparation was carried out through the suspension of ~0.5 mg nanosorbent in 3 ml of ethanol in an ultrasonic bath and then a drop of this solution was put on to a carbon covered 400-mesh copper grid. More than 300 particles were analyzed to get a particle size distribution. Finally, evaporation of the solvent was performed at room temperature. X-ray diffraction (XRD) was performed using a Panalytical Empyrean diffractometer with Ultima + theta-theta high resolution goniometer, the X-ray generator (Cu K radiation, λ = 1.54056Å) with operation conditions at 45 kV and 40 mA. A Specs spectrometer was used for X-ray photoelectron spectroscopy (XPS) measurements using

K lines of Mg (1253.6 eV, 10 mA) as an X-ray source. All lines were referenced to the C 1s line at 284.6 eV. Peak fittings were performed using a Gaussian function. The surface compositions and chemical oxidation states of Pt and Ni have been investigated by using X-ray photoelectron spectroscopy (XPS). For this purpose, Pt 4f and Ni 2p region of the spectrum was evaluated by Gaussian-Lorentzian method and the estimation of the relative intensity of the species have been performed by counting the integral of each peak, after smoothing, subtraction of the Shirley-shaped background. In the XPS spectrum, accurate binding energies (±0.3 eV) have been determined by referencing to the C 1s peak at 284.6 eV.

2.2   Preparation of Pt-Ni@PANI NPs

Pt-Ni@PANI NPs nanosorbents have been prepared by microwave assistance procedure by using an ethylene glycolic (EG) solution of $PtCl_4$ and $NiCl_2$. For this purpose, firstly, $NiCl_2$ (0.25 mmol) and 0.25 mmol of $PtCl_4$ were dispersed in 30 ml of ethylene glycol (EG), and then a stable suspension was obtained via strong stirring. pH of this solution was set to 12 by the NaOH–EG solution and then kept in the central point of a microwave oven (1200 W) for 60 s where EG acted as the reducing agent for $PtCl_4$ and $NiCl_2$ reduction. Finally, the product solution was filtered and then washed with acetone and deionized water. The product was dried under vacuum at 60°C and then they were mixed with PANI (1:1). Pt@PANI NPs and Ni@PANI NPs have also been synthesized by the same procedure to be able to see the effect of the second metal on MB adsorption.

2.3   Adsorption experiments

Firstly, a calibration curve was built for MB solutions in water at different concentrations (2.5, 5, 10, 20, and 30 mg/L). To run the batch adsorption experiments, firstly, 25 mg of the nanocomposite was dispersed in water by using an ultrasonic bath for 2 h. Next, the mixture was mixed with 25 mL of MB solution (30 mg/L) and shaken in a water bath (120 rpm) for 24 hours. At the end of 24 h shaking, pH was adjusted to 5.8 using NaOH and HCl solutions. Dye adsorption experiments were performed in round bottom flasks at room temperature. After separating the nanocomposite particles by centrifugation (4000 rpm for 10 min), the supernatant solution was analyzed to measure the absorbance at 664 nm, which is the absorption band of MB in water, by using a UV-Vis spectrophotometer. Using the calibration curve and the absorbance data, the amount of dye adsorbed was calculated using the following equation:

$$q_e = (C_o - C_e) \, V \, / \, m$$

In this equation, q_e, C_o, C_e, V and m represent the concentration of dye adsorbed (mg/g), initial concentration of dye (mg/L), equilibrium concentration of dye (mg/L), mass of the NPs (g), and volume of solution (L), respectively.

2.4   Testing the reusability of Pt-Ni@PANI NPs

For the reusability of the Pt-Ni@PANI NPs synthesized in this study for MB removal, 15 mg of the nanocomposite was mixed with 25 mL of MB solution (30 mg/L). Next, the mixture was sonicated for 30 min at room temperature. After separating the nanocomposite from the mixture by centrifugation, the supernatant was kept for the spectroscopic analyses. Afterward, for desorption, used nanocomposite was washed with 25 mL of ethanol three times at room temperature and then collected by centrifugation. This washed nanocomposite was reused for a next MB adsorption experiment as described above. To figure out the reusability, those experiments were repeated six times.

3.   Results and discussion

Characterizations of the newly synthesized Pt-Ni@PANI NPs nanoparticles were performed by XRD, TEM, HRTEM and XPS. The XRD patterns of these nanoparticles were displayed in Fig. 1a. The XRD data showed distinct diffraction patterns of Pt-Ni@PANI NPs. The peak at around 25.9° is attributed to the PANI. Furthermore, Pt (111), (200), (220) and (311) planes of the face-centered cubic crystal lattice of Pt-Ni were observed as diffraction peaks at 2θ = 39.82° , 46.02° , 67.62° and 81.40° (JCPDS-ICDD, Card No. 04-802), which points out that Pt-Ni@PANI NPs is in the face-centered cubic crystalline structure. It was shown that Pt-Ni@PANI NPs had a lattice parameter value of 3.887 Å. To do this, the following equation was used by the help of Pt (220) diffraction peak of a prepared catalyst which is a bit smaller than 3.890 Å for pure Pt [42-51] which also indicates the alloy formation. The Debye-Scherrer equation was utilized to calculate the average crystallite particle size [51-55]:

$$d (\text{Å}) = k\lambda/\beta\cos\theta$$

where, k is a coefficient (0.9), λ is the wavelength of X-ray used (1.54056 Å), β is the full width half-maximum of respective diffraction peak (rad), and θ is the angle at the position of peak maximum (rad). Average crystallite Pt particle size of Pt-Ni@PANI NPs was calculated as 3.96 ± 0.34 nm. Moreover, the right shift observed in the XRD patterns of bimetallic Pt-Ni@PANI NPs compared to monometallic Pt one indicates the alloy formation between the metals.

The morphology and structure of the newly synthesized Pt-Ni@PANI NPs were also analyzed by a TEM and HRTEM, and the results were depicted in Fig. 1b, which is in good agreement with the XRD results. The average particle size distribution for the nanoparticles is displayed in Fig. 1b and shows that the particles were mostly spherical, and besides, there was no agglomeration. Furthermore, for Pt-Ni@PANI NPs, the representative atomic lattice fringes were also depicted in Fig. 1b. It was found that, on the prepared catalyst, Pt (111) plane spacing is 0.22 nm, and this is a bit smaller compared to nominal Pt (111) spacing of 0.23 nm. This also indicates the alloy formation of Pt-Ni@PANI NPs.

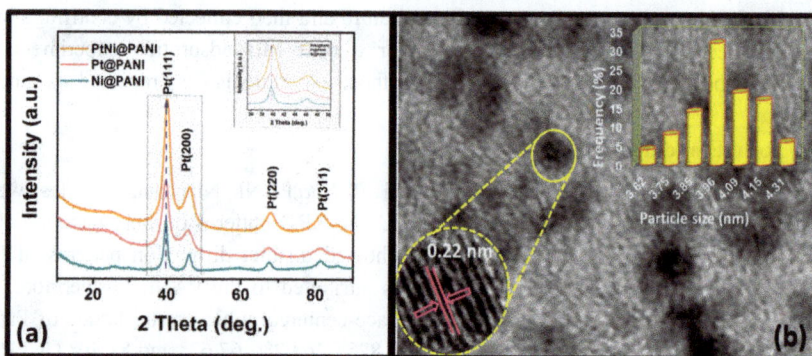

Fig. 1 (a) The XRD pattern of the Pt-Ni@PANI NPs.(b) TEM and HRTEM images of Pt-Ni@PANI NPs.

X-ray photoelectron spectroscopy (XPS) was also used in the present study in order to understand how surface oxidation states of Pt and Ni affected the performance of monodisperse Pt-Ni@PANI NPs. To do that, analyses of the Pt 4f and Ni 2p regions of the spectrum were accomplished. The Gaussian-Lorentzian method [56] was used to arrange the fittings of XPS peaks. To estimate the relative intensities of the species, firstly, smoothing and subtraction of the Shirley-shaped background were done for the peaks gathered in XPS, and then the integral of each peak was calculated. By referencing to the C 1s peak at 284.6 eV, show that all the Pt species in Pt-Ni@PANI NPs, the core level of Pt 4f was at 70.5 eV, indicating that the most Pt atoms were present in the metallic state. The Pt (II) peak at 72.9 eV in Fig. 2a may be caused by the surface oxidation and/or chemisorption of environmental oxygen during the preparation process.

[57-58] Besides, most of the Ni atoms were present in the metallic state at 853.8 eV. The Ni (II) peak at 858.7 eV in Fig. 2b may also be caused by the surface oxidation during the preparation process.

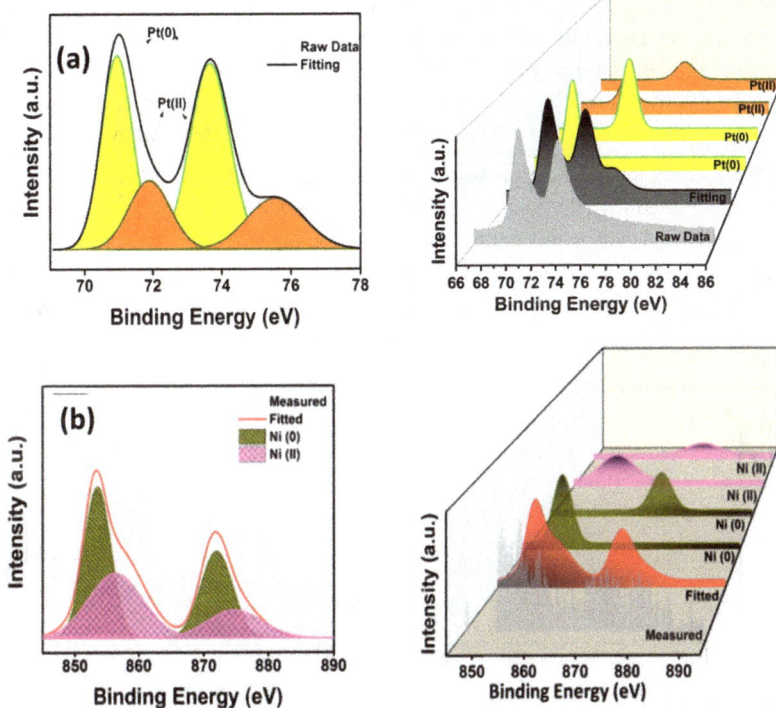

Fig. 2 (a) Pt and (b) Ni XPS spectra results for the Pt-Ni @PANI NPs

Generally, in the literature, bimetallic NPs have been used in alcohol oxidation, dehydrogenation and hydrolysis etc. reactions [59-64,83], however, no study has been performed for the methylene blue (MB) removal from aqueous solutions. On the other hand, as the microwave assisted methodology employed within this work is an appropriate, attractive, rapid, safe, simple and useful method; the synthesis procedure becomes more effective for such a system. For this purpose, after the full characterization was performed, adsorption properties of the Pt-Ni@PANI NPs, Pt@PANI NPs and Ni@PANI NPs were tested for removal of MB from water. For higher MB concentrations, a little dye aggregation was observed and therefore, 10 mg/L Mb solution was utilized to investigate the relationship between the dye adsorption and contact time at room temperature. The results were presented in Fig. 3. Our findings presented that MB adsorption by Pt-Ni@PANI NPs, Pt@PANI NPs and Ni@PANI NPs reached equilibrium after 55 min. This is a short equilibrium time, hence, it could be said that monodisperse Pt-Ni@PANI NPs is efficient adsorbents for MB removal.

In addition, when the contact time was between 0 and 55 min the MB removal efficiency changed (increased with the contact time) significantly. However when it reached 55 min, MB removal efficiency started not to change as it can be seen in Fig. 3b as a linear curve. The possible answer for this type of behavior can be the decreasing MB concentration during the process. For this aim, a calibration curve was made for 2.5, 5, 10, 20, and 30 mg/L MB solutions (Fig. 3c).

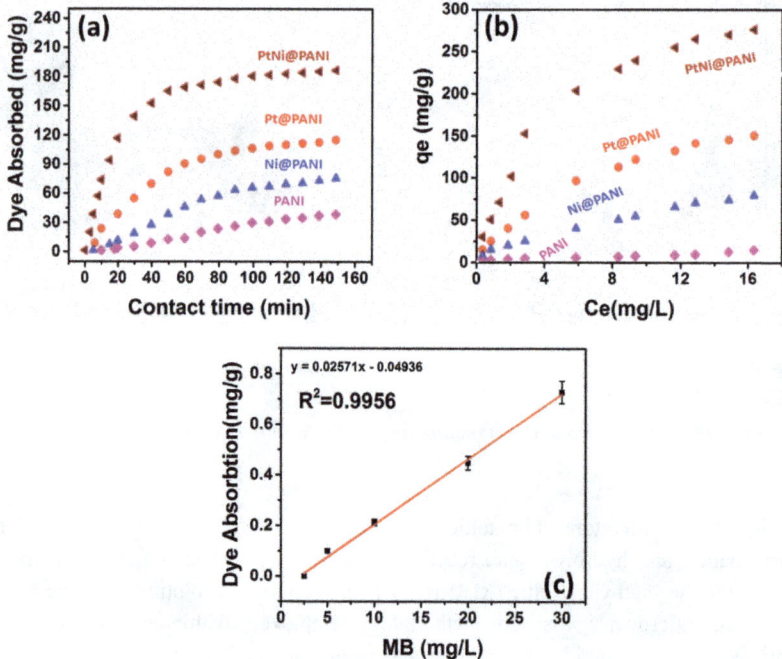

Fig. 3 (a) The relationship between adsorption capacity of Pt-Ni@PANI NPs and various contact time when initial MB concentration was 10 mg/L. (b) Isotherm of the MB adsorption of Pt-Ni@PANI NPs (c) Calibration curve for the solutions having different concentrations of methylene blue.

The adsorption isotherm (with the existence of the Pt-Ni@PANI NPs, Pt@PANI NPs and Ni@PANI NPs) was displayed in Fig. 3b. In this figure, the relationship (at equilibrium) between the amount of Pt-Ni@PANI NPs, Pt@PANI NPs and Ni@PANI NPs adsorbed per unit weight of the dye (qe, mg/g) and dye concentrations (C_e, mg/L) was given. The results showed that the higher adsorption capacity was observed in Pt-Ni@PANI NPs as 271.15 mg/g compared to the Pt@PANI NPs and Ni@PANI NPs. When this finding is compared to previously reported adsorbents [1,2,65-82], the higher results were obtained (Table 1). The possible reasons for this might be (i) the monodispersity of Pt-Ni@PANI NPs; (ii) higher specific surface area and metal contents and (iii) electrostatic interactions between PANI and MB. Additionally, highly stable and reusable nanosorbents were produced. Since the stability of the nanocomposites is very crucial for their practical application as an adsorbent in wastewater treatment processes, the stability of Pt-Ni@PANI NPs was also tested in this study.

Table 1. Adsorption capacities of different materials for MB removal.

Adsorbent	Adsorption capacity (mg MB/g)	References
Pt-Ni@PANI NPs	271.15	This work
Pt@PANI NPs	148.2	This work
Ni@PANI NPs	78.3	This work
PANI	15.2	This work
GO–Fe_3O_4 hybrids	172.6	[1]
PZS nanospheres	20	[2]
MPB-AC	163.3	[65]
MWCNTs with Fe_2O_3	42.3	[66]
Na-ghassoulite	135	[67]
GO	17.3	[69]

GO-Fe$_3$O$_4$	190.14	[70]
Graphene	153.85	[71]
GO-Fe$_3$O$_4$-SiO$_2$	111.1	[72]
MB-wheat straw	274.1	[74]
MB-cotton stalk	147.1	[75]
MB-cucumber peels	111.1	[76]
MB-rice hull ash	17.1	[77]
MB-shaddock peel	309.6	[78]
MB-cottonseed hull	185.2	[79]
MB-banana leaves	109.9	[80]
MB-Bacillus subtilis	169.5	[81]
MB-citrus limetta	227.3	[82]

On the other hand, a good adsorbent material does not only possess high adsorption capability but also perfect desorption property [39]. Therefore, the reusability of Pt-Ni@PANI NPs was tested in our study. For this aim, 6 successive cycles of adsorption-desorption were done. The results are displayed in Fig. 4. Although adsorption capacity of Pt-Ni@PANI NPs for MB removal showed a little decrease for each adsorption-desorption cycle, they still had 55.6 % of the initial efficiency after 6 cycles. Hence, our results showed that the regenerated Pt-Ni@PANI NPs can be used repeatedly as efficient adsorbents for MB removal (Fig. 4). Based on our results, the Pt-Ni@PANI NPs are efficient for MB dye removal from aqueous solutions, have reusability with high adsorption capacity and adsorption rate.

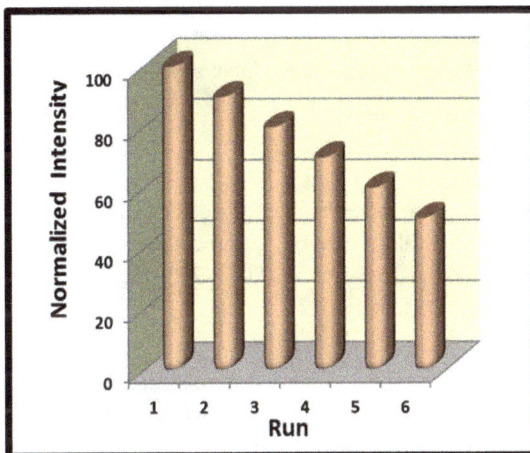

Fig. 4 The reusability graph of the Pt-Ni@PANI NPs for the MB removal. Here, the concentrations of Pt-Ni@PANI NPs and MB were 0.25 g/L and 10 mg/L, respectively. The experiments were done at 25°C and for 30 min contact time.

Conclusions

In summary, an effective, simple and practical Pt-Ni@PANI NPs production method was presented successfully. Our method contains lots of advantageous such as short reaction times, great yields, facile methodology steps and simple work up. Pt-Ni@PANI NPs indicated remarkable nanosorbent performance together with very high dye removal capacity (271.15 mg MB /g nanocomposite) for methylene blue in the water. The possible explanations are: (i) the monodispersity of Pt-Ni@PANI NPs; (ii) high specific surface area; (iii) metal contents; and (iv) electrostatic interactions between PANI and MB, which induced the MB adsorption on Pt-Ni@PANI NPs. Additionally, highly stable and reusable absorbents were produced. It was found that 55.6 % of the initial capacity remains after six adsorption-desorption cycles were performed. Pt-Ni@PANI NPs has promising potential applications for removing organic dyes from polluted waters. This facile, straightforward, and controllable method offers a new pathway for the preparation of new materials with high adsorbent performances, which can find extensive applications. The porous structures provide a large surface area and distance for the dye removal applications.

Acknowledgements

This research was supported by Dumlupinar University Research Funding Agency (2014-05 and 2015-35).

References

[1]   K. Meral, and O. Metin Turk, Graphene Oxide-Magnetite Nanocomposite as an Efficient and Magnetically Separable Adsorbent for Methylene Blue Removal from Aqueous Solution , J. Chem., 38(2014) 775–782. https://doi.org/10.3906/kim-1312-28

[2]   Z. Chen, J. Fu, M. Wang, X. Wang, J. Zhang, Q. Xu, and R. A. Lemons, Self-assembly fabrication of microencapsulated n-octadecane with natural silk fibroin shell for thermal-regulating textiles, Appl. Surf. Sci., 289 (2014) 495–501. https://doi.org/10.1016/j.apsusc.2013.11.022

[3]   B. Yu, X. Zhang, J. Xie, R. Wu, X. Liu, H. Li, F. Chen, H. Yang, Z. Ming, and S.-T. Magnetic graphene sponge for the removal of methylene blue., Appl. Surf. Sci., 2015, 351, 765–771. https://doi.org/10.1016/j.apsusc.2015.05.185

[4]   G. Crini, Non-conventional low-cost adsorbents for dye removal: a review. Bioresour Technol., 97 (2006) 1061–1085. https://doi.org/10.1016/j.biortech.2005.05.001

[5]   M. Rafatullah, O. Sulaiman, R. Hashim, and A. Ahmad, Adsorption of methylene blue on low-cost adsorbents: a review, J. Hazard. Mater., 177 (2010) 70–80. https://doi.org/10.1016/j.jhazmat.2009.12.047

[6]   F. Liu, H. Zou, J. Hu, H. Liu, J. Peng, Y. Chen, F. Lu, and Y. Huo, Fast removal of methylene blue from aqueous solution using porous soy protein isolate based composite beads, Chem. Eng. J., 287 (2016) 410–418. https://doi.org/10.1016/j.cej.2015.11.041

[7]   V.K. Gupta, Suhas, Application of low-cost adsorbents for dye removal--a review. J. Environ. Manage. 90 (2009) 2313–2342. https://doi.org/10.1016/j.jenvman.2008.11.017

[8]   P.N. Bhavani, A.V.R. Krishna Rao, G. Anshu, K. Ashok, K.D. Raj, Synthesis, characterization and enhanced photocatalytic degradation efficiency of Se doped ZnO nanoparticles using trypan blue as a model dye, Appl. Catal. A -Gen. 2013, 459, 106–113. https://doi.org/10.1016/j.apcata.2013.04.001

[9]   L. Bai, Z. Li, Y. Zhang, T. Wang, R. Lu, W. Zhou, H. Gao, and S. Zhang, Synthesis of water-dispersible graphene-modified magnetic polypyrrole nanocomposite and its ability to efficiently adsorb methylene blue from aqueous solution, Chem. Eng. J., 279 (2015) 757–766. https://doi.org/10.1016/j.cej.2015.05.068

[10]   L. Fan, C. Luo, M. Sun, H. Qiu, and X. Li, Synthesis of magnetic β-cyclodextrin-chitosan/graphene oxide as nanoadsorbent and its application in dye adsorption and removal, Coll. Surf, B Biointerfaces, 103 (2013) 601–607. https://doi.org/10.1016/j.colsurfb.2012.11.023

[11]   Z. Dasdelen, Y. Yıldız, S. Eriş, F. Şen, Enhanced electrocatalytic activity and durability of Pt nanoparticles decorated on GO-PVP hybride material for methanol oxidation reaction, Applied Catalysis B: Environmental 219C (2017) pp. 511-516 https://doi.org/10.1016/j.apcatb.2017.08.014

[12]   S. Bozkurt, B. Tosun, B. Sen, S. Akocak, A. Savk, MF. Ebeoğlugil, F. Sen, A hydrogen peroxide sensor based on TNM functionalized reduced graphene oxide grafted with highly monodisperse Pd nanoparticles, Analytica Chimica Acta, 10.1016/j.aca.2017.07.051 https://doi.org/10.1016/j.aca.2017.07.051

[13]   E. Demir, B.Sen, F. Sen, Highly efficient nanoparticles and f-MWCNT nanocomposites based counter electrodes for dye-sensitized solar cells, Nano-Structures & Nano-Objects, 11 (2017) 39-45 https://doi.org/10.1016/j.nanoso.2017.06.003

[14]   G. Baskaya, Y. Yıldız, A. Savk, TO. Okyay, S. Eris, F.Sen, Rapid, Sensitive, and Reusable Detection of Glucose by Highly Monodisperse Nickel nanoparticles decorated functionalized multi-walled carbon nanotubes, Biosensors and Bioelectronics, 91 (2017) 728–733 https://doi.org/10.1016/j.bios.2017.01.045

[15]   B. Sen, S. Kuzu, E. Demir, E. Yıldırır, F. Sen, Highly Efficient Catalytic Dehydrogenation of Dimethly Ammonia Borane via Monodisperse Palladium-Nickel Alloy Nanoparticles Assembled on PEDOT, International Journal of Hydrogen Energy, 10.1016/j.ijhydene.2017.05.115 https://doi.org/10.1016/j.ijhydene.2017.05.115

[16]   S.Akocak, B. Şen, N. Lolak, A. Şavk, M. Koca, S. Kuzu, F. Şen, One-pot three-component synthesis of 2-Amino-4H-Chromene derivatives by using monodisperse Pd nanomaterials anchored graphene oxide as highly efficient and recyclable catalyst, Nano-Structures & Nano-Objects (Invited), 11 (2017) 25–31 https://doi.org/10.1016/j.nanoso.2017.06.002

[17]   Y. Yıldız, S. Kuzu, B. Sen, A. Savk, S. Akocak, F. Şen, Different ligand based monodispersed metal nanoparticles decorated with rGO as highly active and reusable catalysts for the methanol oxidation, International Journal of Hydrogen Energy, 42 (18) (2017)13061-13069.
https://doi.org/10.1016/j.ijhydene.2017.03.230

[18]   R. Ayranci, G.Baskaya, M. Guzel, S.Bozkurt, M. Ak, A.Savk, F. Sen, Carbon based Nanomaterials for High Performance Optoelectrochemical Systems, Chemistry Select, 2 (4)(2017) 1548-1555 https://doi.org/10.1002/slct.201601632

[19]   B.Sen, S. Kuzu, E. Demir, S. Akocak, F. Sen, Monodisperse Palladium-Nickel Alloy Nanoparticles Assembled on Graphene Oxide with the High Catalytic Activity and Reusability in the Dehydrogenation of Dimethylamine-Borane, International Journal of Hydrogen Energy, 10.1016/j.ijhydene.2017.05.113
https://doi.org/10.1016/j.ijhydene.2017.05.113

[20]   B. Sen, S.Kuzu, E. Demir, S. Akocak, F. Sen, Polymer-Graphene hybride decorated Pt Nanoparticles as highly eficient and reusable catalyst for the Dehydrogenation of Dimethylamine-borane at room temperature, International Journal of Hydrogen Energy, 10.1016/j.ijhydene.2017.05.112
https://doi.org/10.1016/j.ijhydene.2017.05.112

[21]   Y.Yildiz, TO. Okyay, B. Sen B. Gezer, S. Kuzu, A. Savk, E. Demir, Z. Dasdelen and F. Sen, Highly Monodisperse Pt/Rh Nanoparticles Confined in the Graphene Oxide for Highly Efficient and Reusable Sorbents for Methylene Blue Removal from Aqueous Solutions, Chemistry Select, 2 (2), 2017, 697-701

[22]   R.Ayranci, G. Baskaya, M. Guzel, S. Bozkurt, M. Ak, A. Savk, F. Sen, Enhanced optical and electrical properties of PEDOT via nanostructured carbon materials: A Comparative investigation, Nano-Structures and Nano-Objects, 11 (2017) 13–19
https://doi.org/10.1016/j.nanoso.2017.05.008

[23]   B.Sen, S. Kuzu, E. Demir, TO. Okyay, F. Sen, Hydrogen liberation from the dehydrocoupling of dimethylamine-borane at room temperature by using novel and highly monodispersed RuPtNi nanocatalysts decorated with graphene oxide, International Journal of Hydrogen Energy, 10.1016/j.ijhydene.2017.04.213
https://doi.org/10.1016/j.ijhydene.2017.04.213

[24]   B. Sen, S. Kuzu, E. Demir, S. Akocak, F. Sen, Highly Monodisperse RuCo Nanoparticles decorated on Functionalized Multiwalled Carbon Nanotube with the Highest Observed Catalytic Activity in the Dehydrogenation of Dimethylamine

Borane, International Journal of Hydrogen Energy, 10.1016/j.ijhydene.2017.06.032 https://doi.org/10.1016/j.ijhydene.2017.06.032

[25] Ö. Karatepe, Y.Yıldız, H. Pamuk, S. Eriş, Z.Dasdelen and F. Şen, Enhanced electro catalytic activity and durability of highly mono disperse Pt@PPy-PANI nanocomposites as a novel catalyst for electro-oxidation of methanol, RSC Advances, 6 (2016) 50851 – 50857 https://doi.org/10.1039/C6RA06210E

[26] B. Gezer, TO Okyay, S. Bozkurt, G. Başkaya, B. Şahin, C. Ulutürk, F.Sen, Reduced Graphene Oxide (rGO) as Highly Effective Material for the Ultrasound Assisted Boric Acid Extraction from Ulexite Ore, Chemical Engineering Research and Design, 117 (2017) 542-548 https://doi.org/10.1016/j.cherd.2016.11.007

[27] Y. Yildiz, E. Erken, H. Pamuk and F. Sen, Monodisperse Pt Nanoparticles Assembled on Reduced Graphene Oxide: Highly Efficient and Reusable Catalyst for Methanol Oxidation and Dehydrocoupling of Dimethylamine-Borane (DMAB), J. Nanosci. Nanotechnol. 16 (2016)5951-5958 https://doi.org/10.1166/jnn.2016.11710

[28] E. Erken, H. Pamuk, Ö. Karatepe, G. Başkaya, H. Sert, O M. Kalfa, F. Şen, New Pt(0) Nanoparticles as Highly Active and Reusable Catalysts in the C1–C3 Alcohol Oxidation and the Room Temperature Dehydrocoupling of Dimethylamine-Borane (DMAB), Journal of Cluster Science, (2016) 27: 9. https://doi.org/10.1007/s10876-015-0892-8

[29] B. Çelik, Y. Yildiz, E. Erken and Y. Koskun, F. Sen, Monodisperse Palladium-Cobalt Alloy Nanoparticles Assembled on Poly (N-vinyl-pyrrolidone) (PVP) as Highly Effective Catalyst for the Dimethylammine Borane (DMAB) dehydrocoupling, RSC Advances, 6 (2016) 24097 – 24102 https://doi.org/10.1039/C6RA00536E

[30] Y. Yıldız, TO. Okyay, B. Sen, B. Gezer, S. Bozkurt, G. Başkaya and F. Sen, Activated Carbon Furnished Monodisperse Pt nanocomposites as a superior adsorbent for methylene blue removal from aqueous solutions, Journal of Nanoscience and Nanotechnology, 17 (2017)4799–4804 https://doi.org/10.1166/jnn.2017.13776

[31] H. Göksu, B. Kilbas and F. Sen, Recent Advances in the Reduction of Nitro Compounds by Heterogenous Catalysts, Current Organic Chemistry (Invited), 21 (9) (2017) 794-820 https://doi.org/10.2174/1385272820666160525123907

[32] H. Göksu, B. Çelik, Y. Yıldız, B. Kılbaş and F. Şen, Superior monodisperse CNT-Supported CoPd (CoPd@CNT) nanoparticles for selective reduction of nitro

compounds to primary amines with NaBH4 in aqueous medium, Chemistry Select, 1 (10) (2016) 2366-2372. https://doi.org/10.1002/slct.201600509

[33] F. Liu, S. Chung, G. Oh, and T. S. Seo, Three-dimensional graphene oxide nanostructure for fast and efficient water-soluble dye removal, ACS Appl. Mater. Interfaces, 4 (2012) 922–927. https://doi.org/10.1021/am201590z

[34] B. Madhumita, M. Arjun, V.V. Srinivasu, S.O. Maurice, Enhanced removal of Cr(VI) from aqueous solution using polypyrrole/Fe3O4 magnetic nanocomposite, J. Hazard. Mater. 190 (2011) 381–390. https://doi.org/10.1016/j.jhazmat.2011.03.062

[35] J. Xu, H. Lv, S. T. Yang, and J. Luo, Preparation of graphene adsorbents and their applications in water purification, Reviews in Inorganic Chemistry, 33 (2013) 139-160. https://doi.org/10.1515/revic-2013-0007

[36] D. R. Dreyer, S. Park, C. W. Bielawski, and R. S. Ruoff, The chemistry of graphene oxide. Chemical Society Reviews, 39 (2010) 228-240. https://doi.org/10.1039/B917103G

[37] F. Ahmed and D. F. Rodrigues, Investigation of acute effects of graphene oxide on wastewater microbial community: a case study. Journal of Hazardous Materials, 256–257 (2013) 33-39. https://doi.org/10.1016/j.jhazmat.2013.03.064

[38] L. Fan, C. Luo, M. Sun, H. Qiu, and X. Li, Synthesis of magnetic β-cyclodextrin–chitosan/graphene oxide as nanoadsorbent and its application in dye adsorption and removal, Colloids and Surfaces B: Biointerfaces, 103 (2013) 601-607. https://doi.org/10.1016/j.colsurfb.2012.11.023

[39] L. Bai, Z. Li, Y. Zhang, T. Wang, R. Lu, W. Zhou, H. Gao, and S. Zhang, Synthesis of water-dispersible graphene-modified magnetic polypyrrole nanocomposite and its ability to efficiently adsorb methylene blue from aqueous solution, Chemical Engineering Journal, 279 (2015) 757-766. https://doi.org/10.1016/j.cej.2015.05.068

[40] C. O'Neill, F. R. Hawkes, D. L. Hawkes, N. D. Lourenco, H. M. Pinheiro, and W. Delee, Colour in textile effluents – sources, measurement, discharge consents and simulation: a review. Journal of Chemical Technology and Biotechnology, 74 (1999) 1009-1018. https://doi.org/10.1002/(SICI)1097-4660(199911)74:11<1009::AID-JCTB153>3.0.CO;2-N

[41] Y. Garsany, A. Epshteyn, A. P. Purdy, K. L. More, and K. E. Swider-Lyons, High-Activity, Durable Oxygen Reduction Electrocatalyst: Nanoscale Composite of

Platinum−Tantalum Oxyphosphate on Vulcan Carbon. J. Phys. Chem. Lett., 2010, 1, 1977–1981. https://doi.org/10.1021/jz100681g

[42] B. Celik, S. Kuzu, E. Erken, H. Sert, Y. Koskun and Fatih Sen Nearly Monodisperse Carbon Nanotube Furnished Nanocatalysts as Highly Efficient and Reusable Catalyst for Dehydrocoupling of DMAB and C1 to C3 Alcohol Oxidation International Journal of Hydrogen Energy, 41 (2016) 3093-3101. https://doi.org/10.1016/j.ijhydene.2015.12.138

[43] H. Goksu, Y. Yıldız, B. Celik, M. Yazici, B. Kilbas and Fatih Sen, Eco-friendly hydrogenation of aromatic aldehyde compounds by tandem dehydrogenation of dimethylamine-borane in the presence of reduced graphene oxide furnished platinum nanocatalyst. Catalysis Science and Technology, 6 (2016) 2318-2324. https://doi.org/10.1039/C5CY01462J

[44] B. Aday, Y. Yıldız, R. Ulus, S. Eris, M. Kaya, and F. Sen, One-Pot, Efficient and Green Synthesis of Acridinedione Derivatives using Highly Monodisperse Platinum Nanoparticles Supported with Reduced Graphene Oxide. New Journal of Chemistry, 40 (2016) 748 – 754. https://doi.org/10.1039/C5NJ02098K

[45] B. Celik, G. Baskaya, H. Sert, O. Karatepe, E. Erken and F. Sen, Monodisperse Pt(0)/DPA@GO nanoparticles as highly active catalysts for alcohol oxidation and dehydrogenation of DMAB. International Journal of Hydrogen Energy, 13 (2016) 5661-5669. https://doi.org/10.1016/j.ijhydene.2016.02.061

[46] B. Celik, Y. Yildiz, H. Sert, E. Erken, Y. Koskun and F. Sen, Monodisperse Palladium-Cobalt Alloy Nanoparticles Assembled on Poly (N-vinyl-pyrrolidone) (PVP) as Highly Effective Catalyst for the Dimethylammine Borane (DMAB) dehydrocoupling. RSC Advances, 6 (2016) 24097 – 24102. https://doi.org/10.1039/C6RA00536E

[47] E. Erken, I. Esirden, M. Kaya and F. Sen, Monodisperse Pt NPs@rGO as Highly Efficient and Reusable Heterogeneous Catalyst for the Synthesis of 5-substituted 1H-tetrazole Derivatives. Catalysis Science and Technology, 5 (2015) 4452. https://doi.org/10.1039/C5CY00864F

[48] H. Pamuk, B. Aday, F. Sen and M. Kaya, Pt Nps@GO as Highly Efficient and Reusable Catalyst for One-Pot Synthesis of Acridinedione Derivatives. RSC Adv, 5 (2015) 49295-49300. https://doi.org/10.1039/C5RA06441D

[49] F. Sen and G. Gokagac, Improving Catalytic Efficiency in the Methanol Oxidation Reaction by Inserting Ru in Face-Centered Cubic Pt Nanoparticles Prepared by a

New Surfactant, tert-Octanethiol. Energy & Fuels, 22 (2008) 1858–1864. https://doi.org/10.1021/ef700575t

[50]   Z. Ozturk, F. Sen, S. Sen and G. Gokagac, The preparation and characterization of nano-sized Pt-Pd alloy catalysts and comparison of their superior catalytic activities for methanol and ethanol oxidation. J. Mater. Sci., 47 (2012) 8134–8144. https://doi.org/10.1007/s10853-012-6709-3

[51]   F. Sen, S. Sen and G. Gokagac, Efficiency enhancement in the methanol/ethanol oxidation reactions on Pt nanoparticles prepared by a new surfactant, 1,1-dimethyl heptanethiol, and surface morphology by AFM. Phys. Chem. Chem. Phys., 13 (2011) 1676-1684. https://doi.org/10.1039/C0CP01212B

[52]   F. Sen and G. Gokagac, Pt Nanoparticles Synthesized with New Surfactans: Improvement in C1-C3 Alcohol Oxidation Catalytic Activity. J. Appl. Electrochem, 44 (2014) 199-207. https://doi.org/10.1007/s10800-013-0631-5

[53]   S. Sen, F. Sen and G. Gokagac, Preparation and characterization of nano-sized Pt–Ru/C catalysts and their superior catalytic activities for methanol and ethanol oxidation. Phys. Chem. Chem. Phys., 13 (2011) 6784-6792. https://doi.org/10.1039/c1cp20064j

[54]   F. Sen and G. Gokagac, The activity of carbon supported platinum nanoparticles towards methanol oxidation reaction – role of metal precursor and a new surfactant, tert-octanethiol. J. Phys. Chem. C, 111 (2007) 1467-1473. https://doi.org/10.1021/jp065809y

[55]   E. Erken, I. Esirden, M. Kaya, F.Sen, A Rapid and Novel Method for the Synthesis of 5-Substituted 1H-tetrazole Catalyzed by Exceptional Reusable Monodisperse Pt NPs@AC under the Microwave Irradiation. RSC Advances, 5 (2015) 68558-68564. https://doi.org/10.1039/C5RA11426H

[56]   Klug, Harold P., and Leroy E. Alexander. "X-ray diffraction methods for polycrystalline and amorphous materials." (1954).

[57]   F. Sen, G. Gökagac, Different sized platinum nanoparticles supported on carbon: An XPS study on these methanol oxidation catalysts. J Phys Chem C, 111 (2007) 5715-5720. https://doi.org/10.1021/jp068381b

[58]   P.T.A. Sumodjo, E.J. Silva, T. Rabochai, Electrosorption of hydroxylated compounds: a comparative study of molecules with three carbon atoms. J Electroanal Chem, 271 (1989) 305. https://doi.org/10.1016/0022-0728(89)80084-1

[59] S. Ertan, F. Sen, S. Sen, G. Gokagac, Platinum nanocatalysts prepared with different surfactants for C1 to C3 alcohol oxidations and their surface morphologies by AFM. J Nanopart Res, 14 (2012) 922-26. https://doi.org/10.1007/s11051-012-0922-5

[60] F. Sen, G. Gokagac, S. Sen, High performance Pt nanoparticles prepared by new surfactants for C1 to C3 alcohol oxidation reactions. J Nanopart. Res, 15 (2013) 1979. https://doi.org/10.1007/s11051-013-1979-5

[61] F. Sen, Y. Karatas, M. Gulcan, M. Zahmakiran, Amylamine stabilized platinum(0) nanoparticles: active and reusable nanocatalyst in the room temperature dehydrogenation of dimethylamine- borane. RSC Adv, 4 (2014) 1526-1531. https://doi.org/10.1039/C3RA43701A

[62] E. Erken, H. Pamuk, Ö. Karatepe, G. Başkaya, H. Sert, O. Murat Kalfa, F. Şen, New Pt(0) Nanoparticles as Highly Active and Reusable Catalysts in the C1–C3 Alcohol Oxidation and the Room Temperature Dehydrocoupling of Dimethylamine-Borane (DMAB). Journal of Cluster Science, 2016, 27: 9. https://doi.org/10.1007/s10876-015-0892-8

[63] T. Orçun, E. Gülin, A. Süheyda, F. Jörgen, W. Ulrika, The treatment of azo dyes found in textile industry wastewater by anaerobic biological method and chemical oxidation. Sep. Purif. Technol., 79 (2011) 26–33. https://doi.org/10.1016/j.seppur.2011.03.007

[64] A.M.S. Mohamad, K.M. Dalia, A.W.A.K. Wan, I. Azni, Cationic and anionic dye adsorption by agricultural solid wastes: A comprehensive review, Desalination 208 (2011) 1–13.

[65] K. T. Wong, N. C. Eu, S. Ibrahim, H. Kim, Y. Yoon, and M. Jang, Recyclable magnetite-loaded palm shell-waste based activated carbon for the effective removal of methylene blue from aqueous solution. Journal of Cleaner Production, 115(2016) 337-342. https://doi.org/10.1016/j.jclepro.2015.12.063

[66] S. Qu, F. Huang, S. Yu, G. Chen, and J. Kong, Magnetic removal of dyes from aqueous solution using multi-walled carbon nanotubes filled with Fe2O3 particles. Journal of Hazardous Materials, 160 (2008) 643-647. https://doi.org/10.1016/j.jhazmat.2008.03.037

[67] Y. El Mouzdahir, A. Elmchaouri, R. Mahboub, A. Gil, and S. Korili, Adsorption of Methylene Blue from Aqueous Solutions on a Moroccan Clay. Journal of Chemical & Engineering Data, 52 (2007) 1621-1625. https://doi.org/10.1021/je700008g

[68]   H. Wang, X. Yuan, Y. Wu, H. Huang, X. Peng, G. Zeng, H. Zhong, J. Liang, M. Ren, Graphene-based materials: fabrication, characterization and application for the decontamination of wastewater and wastegas and hydrogen storage/generation.Advances in Colloid and Interface Science, 195–196 (2013) 19-40. https://doi.org/10.1016/j.cis.2013.03.009

[69]   GK. Ramesha, AV. Kumara, H.B Muralidhara, S. Sampath, Graphene and graphene oxide as effective adsorbents toward anionic and cationic dyes. Journal of Colloid Interface Science, 361 (2011) 270. https://doi.org/10.1016/j.jcis.2011.05.050

[70]   F. He, JT. Fan, D. Ma, L. Zhang, C. Leung, HL. Chan. The attachment of Fe3O4 nanoparticles to graphene oxide by covalent bonding. Carbon, 48 (2010) 3139. https://doi.org/10.1016/j.carbon.2010.04.052

[71]   T. Liu, Y. Li, Q. Du, J. Sun, Y. Jiao, G. Yang, et al. Adsorption of methylene blue from aqueous solution by graphene. Colloids Surf B, 90 (2012) 197. https://doi.org/10.1016/j.colsurfb.2011.10.019

[72]   Y. Yao, S. Miao, S. Yu, LP. Ma, H. Sun, S.J. Wang, Fabrication of Fe3O4/SiO2 core/shell nanoparticles attached to graphene oxide and its use as an adsorbent. Colloid Interface Sci, 379 (2012) 20. https://doi.org/10.1016/j.jcis.2012.04.030

[73]   W. Yao, T. Ni, S. Chen, H. Li, and Y. Lu, Graphene/Fe3O4@polypyrrole nanocomposites as a synergistic adsorbent for Cr(VI) ion removal. Composites Science and Technology, 99 (2014) 15-22. https://doi.org/10.1016/j.compscitech.2014.05.007

[74]   W. Zhang, H. Yan, H. Li, Z. Jiang, L. Dong, X. Kan, et al. Removal of dyes from aqueous solutions by straw based adsorbents: Batch and column studies. Chem Eng J, 168 (2011) 1120–7 . https://doi.org/10.1016/j.cej.2011.01.094

[75]   H. Deng, J. Lu, G. Li, G. Zhang, X. Wang, Adsorption of methylene blue on adsor- bent materials produced from cotton stalk. Chem Eng J 172 (2011) 326–34 . https://doi.org/10.1016/j.cej.2011.06.013

[76]   G. Akkaya, F. Guzel, Application of some domestic wastes as new low-cost biosorbents for removal of methylene blue: kinetic and equilibrium studies. Chem Eng Commun 201(2014)557–78 . https://doi.org/10.1080/00986445.2013.780166

[77]   X. Chen, S. Lv, S. Liu, P. Zhang, A. Zhang, J. Sun, et al. Adsorption of methylene blue by rice hull ash. Sep Sci Technol 47(2012)147–56 . https://doi.org/10.1080/01496395.2011.606865

[78]   J. Liang, J. Wu, P. Li, X. Wang, B. Yang, Shaddock peel as a novel low-cost adsor- bent for removal of methylene blue from dye wastewater. Desalination Water Treat 39(2012) 70–5 . https://doi.org/10.1080/19443994.2012.669160

[79]   Q. Zhou, W. Gong, C. Xie, X. Yuan, Y. Li, C. Bai, et al. Biosorption of methylene blue from aqueous solution on spent cottonseed hull substrate for pleurotus ostreatus cultivation. Desalination Water Treat 29(2011)317–25 . https://doi.org/10.5004/dwt.2011.2238

[80]   RR. Krishni, KY. Foo, BH. Hameed, Adsorptive removal of methylene blue using the natural adsorbent-banana leaves. Desalination Water Treat 52(2014)6104–12 . https://doi.org/10.1080/19443994.2013.815687

[81]   A. Ayla, A. Cavus, Y. Bulut, Z. Baysal, C. Aytekin, Removal of methylene blue from aqueous solutions onto Bacillus subtilis : determination of kinetic and equilib- rium parameters. Desalination 51 (2013)7596–603 . https://doi.org/10.1080/19443994.2013.791780

[82]   S. Shakoor, A. Nasar, Removal of methylene blue dye from artificially contaminated water using citrus limetta peel waste as a very low cost adsorbent, Journal of the Taiwan Institute of Chemical Engineers, 66 (2016) 154–163 https://doi.org/10.1016/j.jtice.2016.06.009

[83]   S. Shakoor, A. Nasar, Adsorptive treatment of hazardous methylene blue dye from artificially contaminated water using cucumis sativus peel waste as a low-cost adsorbent, Groundwater for Sustainable Development, 5 (2017) 152–159 https://doi.org/10.1016/j.gsd.2017.06.005

Chapter 5

A Basic Overview of Fuel Cells: Materials and Applications

Jayshree Ramkumar* and S. Chandramouleeswaran

Analytical Chemistry Division, Bhabha Atomic Research Centre, Mumbai - 400085, India

* jrk@barc.gov.in

Abstract

In this chapter, the history of fuel cells from invention to modern days is discussed. The 19th century saw the development of steam engine while in the 20th century, the combustion engine was dominant and it is quite probable that the 21st century will be the era of fuel cells. The various types of fuel cells with their advantages, disadvantages and principal applications are reviewed in this chapter.

Keywords

Fuel Cells, Principle of Operation, History of Fuel Cells, Fuel Design

Contents

1.   Introduction

Fuel cells are paramount in various applications of providing power [1]. Though the applications of fuel cells have increased exponentially in the last two decades, fuel cells themselves are not a new concept. The history of fuel cells dates back about two centuries. Before proceeding further, it will be essential to understand the basics of a fuel cell.

A fuel cell is a device that produces electricity through a chemical reaction involving a fuel (compound that can be oxidized may be hydrogen, methane, propane, methanol, diesel fuel, gasoline etc.) and an oxidant with water being the main by-product and small amount of nitrous oxide if air is used as the oxidizer [2-4]. Fuel cells differ from batteries [5] since former requires a continuous source of fuel and oxidant to sustain the chemical reaction, whereas in a battery the chemicals present in the battery react with each other to generate an electromotive force (emf). Fuel cells can produce electricity continuously for as long as these inputs are supplied. Fuel cells can be used to power just about anything conceivable, from cars to mobile phones to space vehicles. Fuel cells offer several advantages over conventional power sources like reduced dependence on fossil fuels, long useful life, high efficiency, relative safety, essentially zero toxicity, minimal maintenance costs, reduced pollution, etc. Fuel cells have advantages of zero emissions with high fuel efficiency which is locally available and do not need to be imported and these are a definite advantage over the gasoline based internal combustion engines (ICE) which give high emissions and also the fuel used is often imported. It can be argued that the disadvantages of gasoline based ICE can be overcome with battery electric vehicles which also do not give emissions or require imported fuel. But these suffer from disadvantages of limited range, long charging time, packaging and performance being affected by weather. Fuel cells overcome the limitations of the battery as they have further advantages of long range, quick refueling, limited weather impact and scalable to a wide range of vehicle sizes.

The efficiency of fuel cells is higher and not limited by Carnot cycle and they do not need to be recharged and can produce continuous power in presence of fuel and oxidant. Mostly all types of fuel cells consist of an electrolyte which allows a charge flow between the anode and cathode. Most of the fuel cell power systems comprise of different components *viz*: unit cells (in which the electrochemical reaction occurs), stacks (in which individual cells are combined by electrically connecting the cells to form units with the desired output capacity), balance of plant which comprises components that provide feed stream conditioning (including a fuel processor if needed), thermal management, and electric power conditioning among other ancillary and interface functions.

A fuel cell does not burn its fuel but oxidation occurs at a temperature much lower than that produced by active combustion [6]. A fuel cell can be recharged by filling a tank or from a continuously available external supply of fuel. A schematic representation is given in Fig 1. In a common form of hydrogen fuel cell, known as the proton exchange membrane (PEM), hydrogen is delivered to a positive electrode called the anode. At the anode, hydrogen atoms are broken down or ionized into their constituent protons and electrons. The protons permeate through an electrolyte membrane to a negative electrode called the cathode. Electrons travel from the cathode to the anode through an external load, which converts the resulting current to useful power. Within the cell, oxygen molecules react with the protons permeating through the electrolyte membrane and the electrons arriving through the external load. The result is water, the principal byproduct of all hydrogen-based energy sources.

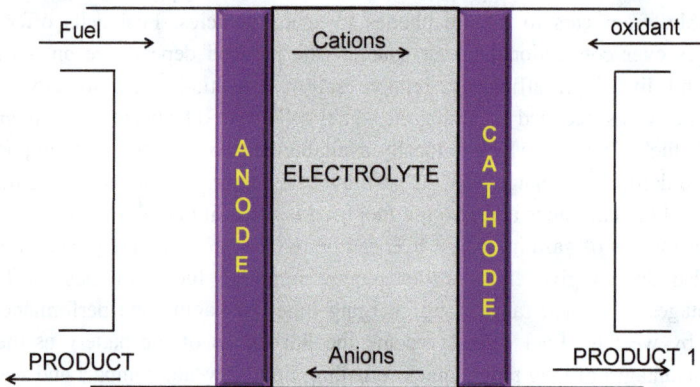

Fig. 1 Schematic representation of principle of fuel cell.

2. History of fuel cells

The progress of fuel cell technologies is a chronicle of implausible pioneering and innovations. Scientists and inventors have made groundbreaking contributions. However, there are still some debates as to who actually discovered the concept of fuel cells. The Department of Energy of the United States [7] believes that it was the German chemist C.F. Schönbein, who in 1838 conducted the first scientific research on the phenomenon of a fuel cell. However, there are contradictory reports in the literature [8] stating that it was Sir W.R. Grove, who pioneered the concept of fuel cell by immersing two platinum electrodes on one end in a solution of sulphuric acid and the other two ends separately sealed in containers of oxygen and hydrogen, a constant current was found to be flowing between the electrodes. The sealed containers contained water together with the respective gases. It was observed that as the current flows, the water level in the two tubes increased. The next step was to apprehend that combining pairs of electrodes connected in series resulted in "a gas battery", i.e. the first fuel cell. Grove made a very imperative observation that the reaction can occur only when there is ample area available for reaction between gas, electrolyte and electrode. Extensive work on fuel cell has been carried out by Grove [9]. The electrolysis of water (splitting into hydrogen and oxygen using electricity) was discovered in 1800 by British scientists Sir A. Carlisle and W. Nicholson and it was acknowledged that they were the first scientists to produce a chemical reaction using electricity. The studies were carried out by connecting one end of a pair of conducting wires to the electrodes of a Volta battery, while the other end was immersed in a saline solution which acts as a conductor. hydrogen and oxygen were collected at the ends of the two electrodes. In 1989, L. Mond along with C. Langer discovered the electrochemical process (Mond process) of purifying nickel [10]. F.W. Ostwald in 1893 experimentally ascertained the consanguinity between electrodes and electrolytes [11] and helped in understanding the unsolved concepts of Grove's gas battery. Mond and Langer were the first to refine Grove's cell by using porous and three-dimensional shaped electrodes. It was also realized by Mond and Langer that coal could be used as a source of hydrogen for fuel cells. This was divergent from the views of Grove, who felt that only pure hydrogen could be used as fuel. Further work was carried out by W.W. Jacques [12] and in 1900, W. Nernst used zirconium as a solid electrolyte. In 1921, Baur constructed the first molten carbonate fuel cell [13] and in 1930's he worked on high temperature system based on solid oxide electrolytes. In 1933, T.R. Bacon developed the first fuel cell that converted air and hydrogen into electricity through an electrochemical process. He further built a high pressure nickel electrode based fuel cell and also one that could be used in submarines and also in the Apollo spacecraft. His research was focussed on the use of non-precious common metals in a non-corrosive environment to improve lifetime and higher efficiency. In 1950, Teflon was available and could be used in fuel cells with platinum or carbon electrodes in presence of acid or alkaline electrolytes respectively. In 1955, T. Grubb a chemist with General Electric Company (GE) modified the original design of the fuel cell by using ion exchange polystyrene sulphonate membrane as the electrolyte. In 1958, L. Niedrach envisioned a method of depositing platinum on the membrane thus enabling its use as a catalyst for the oxidation and reduction reactions of hydrogen and oxygen respectively. The technology for both NASA and McDonnell Aircraft were developed by GE during the Gemini program. In 1959, H. Ihrig and his team built a fuel cell with 15kW for a tractor from Allis-Chalmers. It consisted of a 20 hp tractor with a fuel cell made of 1008 cells of 1 V per cell and KOH electrolyte. It used a mixture of gases like propane and compressed gases as fuel and oxygen as oxidizing agent [14]. In the 1960's, the main focus was on acid electrolyte fuel cell and catalyst platinum in two different ways. One of them used a polymeric electrolyte that is simple and reliable. The other form was developed to use directly the fuels derived from the coal thus making it possible to work at temperatures of 150-200°C. G.H.J. Broers and J.A.A. Ketelaar used a mixture of lithium carbonate, sodium and/or potassium impregnated on magnesia sintered porous disk which can be operated at a very high temperature of 650°C. In 1961, G.V. Elmore and H.A. Tanner made an intermediate temperature phosphoric acid fuel cell [15] using a mixture of phosphoric acid and silicon dust (35%:65%) wedged to Teflon. The observations of these studies showed that electrochemical reduction does not occur during the fuel cell operation and fuel cell operation can be carried out in the air instead of pure oxygen. The authors stated that the fuel cell was functional for a period of six months at 90 mA/cm^2 and 0.25 V devoid of any perceptible attrition. The high temperature of operation of fuel cells was achieved in 1962 by J. Weissbarf and R. Ruka [16] using ceramic oxide impregnated with zirconia while in 1965 research using molten carbonate fuel cells were evaluated for their performance. Fuel cells developed since 1970 have a greater area of action, longer lifetime and increased performance efficiency. The efficiency of phosphoric acid, solid oxide and molten carbonate fuel cells were 45, 50 and 60% respectively. In the present day, fuel cells with direct application in the automotive industry are manufactured. In most of the practical fuel cell applications, unit cells are combined in a modular way into a cell stack to achieve the required voltage and power output level. Generally, the stacking involves connecting of more than one unit cells in series via an electrically conductive interconnection. The different stacking arrangements are like planar-bipolar stacking and stack with tubular cells.

3.   Types of fuel cell designs

Some fuel cells operations are the reverse of electrolysis. There are a large variety of fuel cells based on different fuels and oxidants. Most fuel cell power systems comprise a number of components (i) unit cells, in which the electrochemical reaction takes place; (ii) stacks, in which individual cells are combined by electrically connecting the cells to form units with the desired output capacity, (iii) balance of plant which comprises components that provide feed stream conditioning (including a fuel processor if needed), thermal management, and electric power conditioning among other ancillary and interface functions. Though there are different types of fuel cells, they work on the same comportment. A fuel cell is made of three adjacent segments, viz., the anode, the electrolyte and the cathode. Two chemical reactions occur at the interfaces of the three segments. The reaction occurs with the consumption of fuel and production of electricity along with generation of water or carbon dioxide. At the anode, the catalyst present oxidizes the fuel (hydrogen) resulting in a cation and free electron. The electrons pass through a wire and produce current while the cations travel through the electrolyte towards the cathode, where it combines with another chemical (oxygen) to produce a different chemical species (water or carbon dioxide). The electrodes in fuel cells are usually porous and conducting carbon-based materials coated with a catalyst (usually nickel).

The fuel cells are classified according to the choice of electrolyte and fuel. The major types of fuel cells are (i) Proton exchange membrane fuel cell (PEMFC) [(a) Direct formic acid fuel cell (DFAFC); (b) Direct Ethanol Fuel Cell (DEFC)], (ii) Alkaline fuel cell (AFC) [(a) Proton ceramic fuel cell (PCFC);(b) Direct borohydride fuel cell (DBFC)] , (iii) Phosphoric acid fuel cell (PAFC), (iv) Molten carbonate fuel cell (MCFC), (v) Solid oxide fuel cell (SOFC). The variations arise due to the difference in the electrolyte and oxidant and fuel. The classifications are further made on the basis of the temperature of operation. The low operating temperature is in the range of 50–250 °C for PEMFC, AFC and PAFC, and high operating temperature in the range of 650–1000 °C like MCFC and SOFC. Fig. 2 gives a schematic representation of the various fuel cells normally encountered.

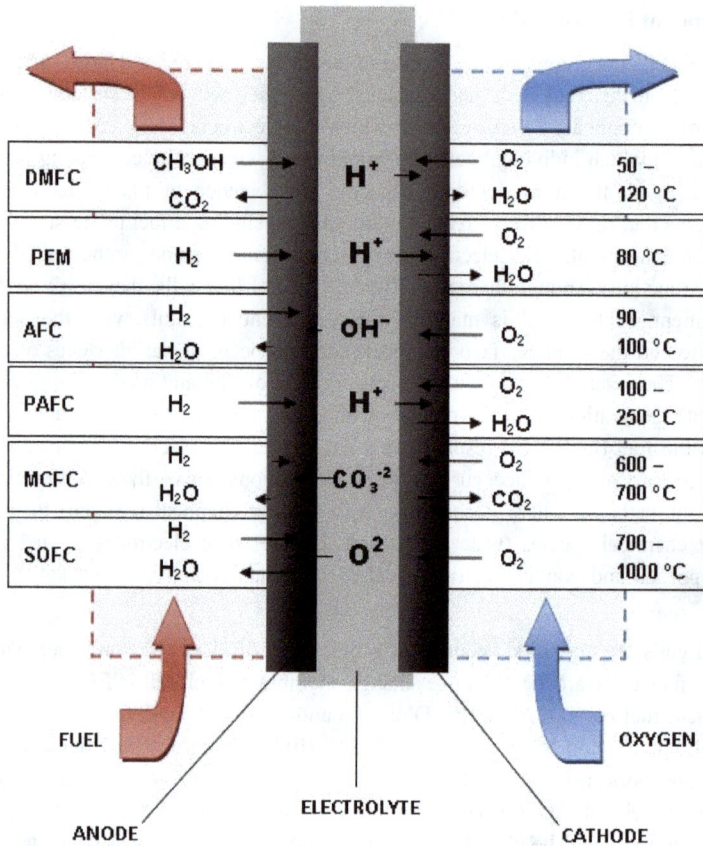

Fig. 2 Schematic representation of various fuel cells.

Some important types of fuel cells are briefly discussed in the following subsections.

3.1   Proton exchange membrane fuel cell (PEMFC)

In this fuel cell, an ion exchange membrane (usually fluorinated sulphonic acid polymer) is used as an electrolyte through which protons are conducted. The problems of corrosion are small. In general, carbon electrodes with platinum electro–catalyst are used for both anode and cathode, and with either carbon or metal interconnection [17]. Water management in the membrane is critical for efficient performance; the membrane should remain hydrated throughout the process thus restraining the fuel cell operation temperature to below 100 °C. It is usually operated in the range of 60-80 °C. H_2 rich gas with minimal or no CO and high catalyst load is required for both the anode and cathode. There is a need for fuel pre-treatment. These are used in automotive and portable applications. The advantages are that a solid electrolyte provides excellent resistance to gas crossover and also high current densities. Also due to low operating temperature, the startup is quite rapid. Due to the narrow temperature range of operation, the main disadvantage is the thermal management of the PEFC. Also, the PEFC is quite sensitive to trace level contaminants including CO, S species and ammonia. A direct formic acid fuel cell is a subcategory of PEFC, wherein the inlet fuel formic acid is fed to the anode. The use of formic acid results in improvement of efficiency as crossover across the polymer is not possible. Also formic acid is safer to use. But the main disadvantage of the system using platinum as a catalyst is high overvoltage which is overcome by using palladium as a catalyst. The direct ethanol fuel cell uses ethanol as input fuel instead of hydrogen. The basic reactions involved are similar to that of PEFC.

Direct-methanol fuel cells or DMFCs are a subcategory of proton-exchange fuel cells in which methanol is used as the fuel. The main advantage of methanol is the easy transportation and an energy-dense yet reasonably stable liquid at all environmental conditions. These cells are having quite low efficiency, so they are very useful especially in portable applications where energy and power density are more important than efficiency.

3.2   Alkaline fuel cells (AFC)

Alkaline Fuel Cell (AFC) also known as the bacon fuel cell. These cells use potassium hydroxide (KOH) or some other alkaline aqueous solution below 100 °C. The electrode reaction takes place by supplying hydrogen (as fuel) to the fuel electrode and oxygen (as an oxidizing agent) to the air electrode. Thus a current flow is obtained [18]. The gases should not have any carbon dioxide to avoid the deterioration of the aqueous electrolyte. At higher concentration of the electrolyte, the water activity is reduced and therefore electrolytic characteristics are improved. However, the reduction of water vapor pressure makes it difficult for the removal of water produced during the reaction. AFCs find extensive applications in transportation. The main advantage of the fuel cell is that it is a well-established technology and is electrically efficient. However, the main disadvantage is its sensitivity to carbon dioxide presence which creates deterioration of the cell electrolyte. Hence stringent precautions are needed to remove the presence of carbon dioxide in the air used. Protonic ceramic fuel cell (PCFC) is a new type of fuel cell developed with ceramic electrolyte. It is operated at 750°C using gaseous hydrocarbons as fuels. The presence of solid electrolyte makes it advantageous as the issues of dehydration of membrane do not arise (in PEFMC) or leak of liquid (PAFC). The main disadvantage is the low current density. This can be overcome by reduction of electrolyte thickness. Direct borohydride fuel cells (DBFC) use sodium borohydride as fuel. The resulting reactions produce borax. This cell can be operated at low temperatures of 70°C. The main advantages of this kind of cell are high power density, the absence of platinum catalyst but the main disadvantage is the low efficiency (35%). Hence ongoing research on the use of inexpensive catalysts to improve efficiency is going on [19]. This fuel cell is in the developmental stage. The hydroxyl ions pass through the electrolyte and complete the circuit and electrical energy is produced. At the anode, hydrogen gas molecules react with hydroxyl ions to produce water and the electrons released in this reaction flow to the cathode through an external circuit and react with water molecules and oxygen to generate hydroxyl anions. AFCs generally perform in temperatures between 60 and 90 °C. AFCs are classified as low operating temperature fuel cells with low-cost catalysts, the most common being nickel. The electrical efficiency of AFCs is about 60% and can generate electricity up to 20 kW. NASA used AFCs to supply drinking water and electric power to the shuttle missions for space applications and presently being used in submarines, boats, forklift trucks etc. AFCs are the most cost efficient type of fuel cells since the electrolyte and catalyst used are low-cost materials (potassium hydroxide and nickel respectively). The design of AFCs is also simple. Since hydrogen and pure oxygen are utilized to produce water, heat and electricity sources and the water is used for drinking purposes in spacecraft and space shuttle fleets. They have no greenhouse gas emissions. Alkaline technology is the longest established technology for both space and submarine applications that is electrically efficient. Despite the various advantages, AFCs suffer from poisoning of water-soluble electrolyte (KOH) by carbon dioxide. KOH absorbs CO_2 resulting in the formation of potassium carbonate (K_2CO_3) and subsequently, the fuel cell is poisoned. The sensitivity of the electrolyte to CO_2 required the use of highly pure H_2 as a fuel or else if ambient air is used as the oxidant, the CO_2 in the air must be removed. Therefore, AFCs typically use purified air or pure oxygen which in turn increases the operating costs. There is a need to find an alternative material to be used as the electrolyte.

3.3   Phosphoric acid fuel cell (PAFC)

In PAFC, the electrolyte consists of concentrated phosphoric acid and silicon carbide which is used to retain the acid. The catalyst is normally Pt or its alloys. The fuel cells are operated in the temperature range of 150-220 °C. At lower temperatures, the ionic conductivity of phosphoric acid is poor while the poisoning of the Pt catalyst by CO becomes a major issue. The porous electrodes are made of a mixture of electrocatalyst supported on carbon black and a polymeric binder which bind the carbon black particles together. A porous carbon paper substrate acts as a strong support for the electrocatalyst layer and as the current collector. The heat generated is removed by liquid or gas coolants which are passed through the cooling channels in the cell stack [20]. The efficiency of PAFC is 40-47 % and finds applications as stationary power generators. One of the main advantages of PAFC is that the quality of DC power produced is excellent. There is also no requirement of highly pure hydrogen and therefore the need for pre-treatment of fuel is not needed. The chances of corrosion are reduced as the electrolyte is distributed within a porous layer of silicon carbide that separates anode and cathode. Fuel cells provide high-quality DC power. The advantages of PAFCs are that impure hydrogen can be used as a fuel and also emission free. Moreover, the problems of corrosion are reduced as the electrolyte is distributed in a porous layer of silicon carbide separating anode and cathode. However, these fuel cells are associated with many disadvantages. The main disadvantages are the larger weight of the PAFC`s and slow start up time due to high operating temperatures coupled with the process being expensive due to use of platinum as catalyst results in limited applications of these PAFCs.

3.4   Molten carbonate fuel cell (MCFC)

MCFCs are high-temperature fuel cell operating at 650 °C and use a molten carbonate salt mixture as the electrolyte. Natural gas is used directly and therefore the operating costs can be reduced. The efficiency is in the range of 60-80%. The electrolyte used is molten carbonate salt mixture. At 650 °C, the mixture melts and allows the conduction of carbonate ions which combine with hydrogen to form water at the cathode. The reaction results in the production of carbon dioxide [21]. The main uses of MCFC are in marine and military applications. The main advantage of MCFC is its low cost. The efficiency of the cells is quite high. Other advantages are the use of non-expensive catalysts and carbon monoxide and even certain hydrocarbons can be used as fuels. Also, the high-temperature waste heat boosts the system efficiency. The disadvantages are the need for high temperature which results in different issues with materials and also the problem of corrosion of electrolyte. They are used in marine and military applications and also in power plants.

3.5   Solid oxide fuel cell (SOFC)

Solid oxide fuel cells use a ceramic material called Yttria stabilized zirconia (YSZ) as the electrolytes. The cells are in the form of tubes and can be operated at temperatures of 800-1000 °C using natural gas as fuel. The conversion of fuel to electricity is carried out with an efficiency of 50-60%. SOFCs are used in a mobile or stationary generation system and as auxiliary power units to run electrical appliances. The electrochemical reaction at the anode is the combining of hydrogen and oxygen producing water and electron which then reaches the cathode to react with oxygen to oxide anion. The overall reaction is the formation of water from hydrogen and oxygen. SOFC has large numbers of applications in the transportation sector and also as auxiliary power units to run electrical systems like air conditioning. The advantages include high efficiency, use of natural gas as fuel, solid electrolytes which avoid the problems of electrolyte movement or flooding in electrodes [22]. The main disadvantage is the high operating temperature which may cause deformities in materials and hinder production as the availability of compatible materials is an issue. The problems of corrosion of the metal stack are also a disadvantage that needs to be tackled. Due to all these issues, the manufacturing cost is a major issue.

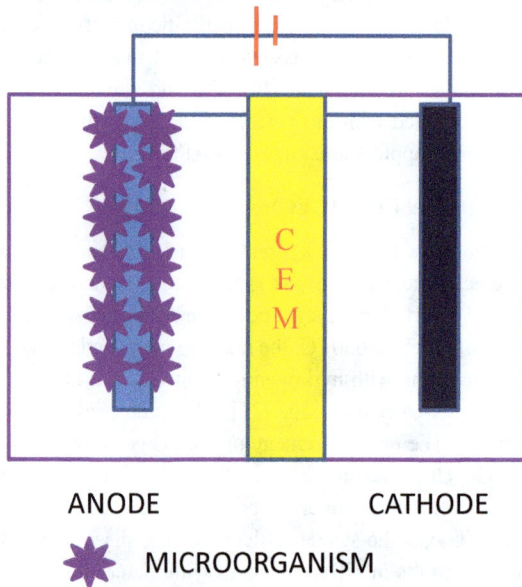

ANODE   CATHODE

MICROORGANISM

Fig. 3 Schematic representation of a microbial fuel cell.

3.6   Microbial fuel cells (MFC)

A microbial fuel cell is a bioreactor that converts the chemical energy of organic reactions to electrical energy using microorganisms as a catalyst in anaerobic conditions. The use of bacteria for generation of electricity is age-old knowledge. It has regained the interest of many researchers due to the ongoing energy quandary and also due to the added advantage of MFC [23,24]. The main advantage of MFC is that the bacteria used to generate electricity can also result in biodegradation of organic matters and wastes. The schematic representation (Fig. 3) shows that the anodic anaerobic and cathodic aerobic chambers separated by a proton exchange membrane. This is a form of PEFC. Microbes along with substrate in the anodic chamber result in the production of carbon dioxide due to oxidation. The anaerobic conditions are maintained to improve the efficiency of microbes. Various kinds of substances can be used as substrates [25].

The various reactions that occur can be understood using a typically used substrate namely acetate. At the anode, acetate anions combine with water to produce carbon dioxide, proton and electrons while at the cathode oxygen combines with protons to produce water.

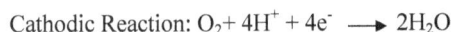

Anodic Reaction: $CH_3COO^- + 2H_2O \longrightarrow 2CO_2 + 7H^+ + 8e^-$

Cathodic Reaction: $O_2 + 4H^+ + 4e^- \longrightarrow 2H_2O$

The net reaction is the breakdown of the substrate to carbon dioxide and water with an associated production of electricity due to the flow of electrons from the anode to the cathode in the external circuits. A large number of microbes can also be used in the MFC [26]. Graphite, graphite felt, carbon paper, carbon cloth, Pt, Pt black, reticulated vitreous carbon (RVC) can be used as anode materials while Graphite, graphite felt, carbon paper, carbon cloth, Pt, Pt black, RVC are used as the cathode. Both anode and cathode are essential components of the fuel cell. Materials like Glass, polycarbonate, Plexiglas can be used for the construction of both anodic and cathodic chambers. Due to their complex design, two-compartments MFCs are difficult to scale-up even though they can be operated in either batch or continuous mode. So an excellent alternative is the one compartment MFCs that offer simpler designs and are economical. In these, only the anodic chamber is present. There are various models reported in the literature [27-32] wherein cathode can be directly exposed to the air, bound to PEM, place outside of a tubular MFC or partitioning of a single chamber.

4. Conclusions

Fuel technology has gained importance due to a growing interest in clean and renewable energy sources. It is expected that the energy sector will see a complete revolutionary change due to the fuel cell technology. With ongoing research on various aspects of fuel cells, this technology may be the core for utilization of various energy generation options.

References

[1]   S. Bhattacharyya (ed.), Rural Electrification Through Decentralised Off-grid Systems in Developing Countries, Green Energy and Technology, Springer Verlag London, 2013, doi: 10.1007/978-1-4471-4673-5_2 https://doi.org/10.1007/978-1-4471-4673-5_2

[2]   U. Bossell, The birth of the Fuel Cell 1835–1845. Power for the 21st century, European Fuel Cell Forum, 2000.

[3]   M. Ball, M. Weeda, The hydrogen economy—Vision or reality? Int. J. Hydr. Energy, 40 (2015) 7903-7919. https://doi.org/10.1016/j.ijhydene.2015.04.032

[4]   B. Cook, An introduction to fuel cells and hydrogen technology. Vancouver, Canada: Heliocentris; 2001.

[5]   M. Winter, R.J. Brodd, What are batteries, fuel cells, and super capacitors? Chem. Rev., 104.(2004) 4245–4270. https://doi.org/10.1021/cr020730k

[6]   R.S. Khurmi, Materials Science, 1987, S. Chand Publishing Co., 1987.

[7]   E.O. Rivera, A.R. Hernandez, A. Febo, Understanding the history of fuel cells. IEEE Conference on the History of Electric Power, 1 (2007) 117–122.

[8]   K. Kordesch, G. Simader, Fuel Cells and their Applications, Weinheim: VCH, 1996, p.38. https://doi.org/10.1002/352760653X

[9]   J. Wisniak, Electrochemistry and fuel cells: the contribution of William Robert Grove, Ind. J. Hist. Sci., 50 (2015) 476-490. https://doi.org/10.16943/ijhs/2015/v50i4/48318

[10]   P.B.L. Chaurasia, Y. Ando, T. Tanaka, Regenerative fuel cell with chemical reactions, Energ. Conv. Manag., 44 (2003) 611–628. https://doi.org/10.1016/S0196-8904(02)00066-3

[11]   A.B. Stambouli, E. Traversa, Solid oxide fuel cells (SOFCs): a review of an environmentally clean and efficient source of energy, Renew. Sustain. Energ. Rev., 6 (2002) 433-455. https://doi.org/10.1016/S1364-0321(02)00014-X

[12]   J. Appleby, From Sir William Grove to today: fuel cells and the future, J. Power Sour., 29 (1990) 3-11. https://doi.org/10.1016/0378-7753(90)80002-U

[13]   C. Stone, A.E. Morrison, From curiosity to power to change the world, Solid State Ion., 152-153 (2002) 1-13. https://doi.org/10.1016/S0167-2738(02)00315-6

[14]   N.S.N.V. Vardhan, G.H.N. Rao, Int. J. Adv. Eng. Globala Technol., 4 (2016) 1221-1225.

[15]   G.V. Elmore and H.A. Tanner, Intermediate Temperature Fuel Cells, J. Electrochem. Soc., 108 (1961) 669-671 https://doi.org/10.1149/1.2428186

[16]   J. Weissbart and R. Ruka, A Solid Electrolyte Fuel Cell, J. Electrochem. Soc., 109 (1962) 723-726. https://doi.org/10.1149/1.2425537

[17]   Y. Wang, K.S. Chen, J. Mishler, S.C. Cho, X.C. Adroher, A review of polymer electrolyte membrane fuel cells: Technology, applications, and needs on fundamental research, App. Ener. 88 (2011) 981-1007. https://doi.org/10.1016/j.apenergy.2010.09.030

[18]   M. Farooque and H.C. Maru, Fuel cells—the clean and efficient power generators, IEEE Proc., 89 (2001) 1819-1829.

[19]   J.B. O'Sullivan, Fuel cells in distributed generation, IEEE Proc., 1 (1999) 568-572.

[20]   M.W. Ellis, M.R.V. Spakovsky, D.J. Nelson, Fuel cell systems: efficient flexible energy conversion for the 21st century. IEEE Proc., 89 (2001) 1808–1818.

[21]   M. Spinelli, M.C. Romano, S. Consonni, S. Campanari, M. Marchi, G. Cinti, Application of molten carbonate fuel cells in cement plants for CO2 capture and clean power generation, Ener. Procedia, 63 (2014) 6517-6526. https://doi.org/10.1016/j.egypro.2014.11.687

[22]   A.J. Jacobson, Materials for solid oxide fuel cells, Chem. Mater., 22 (2010) 660-674. https://doi.org/10.1021/cm902640j

[23]   A.M. Khan, Electricity Generation by Microbial Fuel Cells, Adv. Nat.App. Sci. 3 (2009) 279-286.

[24]   P.K. Barua and D. Deka, electricity generation from bio waste based microbial fuel cells, Int. J. Ener. Infor. Comm., 1 (2010) 77-92.

[25]   D. Pant, G.V. Bogaert, L. Diels, K. Vanbroekhoven, A review of the substrates used in microbial fuel cells (MFCs) for sustainable energy production, Bioreso. Technol., 101 (2010) 1533-1543. https://doi.org/10.1016/j.biortech.2009.10.017

[26]   Z. Du, H. Li and T. Gu, A state of the art review on microbial fuel cells: A promising technology for wastewater treatment and bioenergy, Biotechnol. Adv., 25 (2007) 464-482. https://doi.org/10.1016/j.biotechadv.2007.05.004

[27] M. Rahimnejad, A. Adhami, S. Darvari, A. Zirepour and S.E. Oh, Microbial fuel cell as new technology for bioelectricity generation: A review, Alex. Eng. J., 54 (2015) 745-756. https://doi.org/10.1016/j.aej.2015.03.031

[28] D.H. Park, Y.K. Park, C.E. So, Application of single-compartment bacterial fuel cell (SCBFC) using modified electrodes with metal ions to wastewater treatment reactor, J. Microbiol. Biotechnol., 14 (2004), 1120-1128.

[29] M. Rahimnejad, A. Adhami, S. Darvari, A. Zirepour, S.E. Oh, Microbial fuel cell as new technology for bioelectricity generation: A review, Alex. Eng. J., 54 (2015) 745-756. https://doi.org/10.1016/j.aej.2015.03.031

[30] J. Chouler, G.A. Padgett, P.J. Cameron, K. Preuss, M.M. Titirici, I. Ieropoulos, M.D. Lorenzo, Towards effective small scale microbial fuel cells for energy generation from urine, Electrochim. Acta, 192 (2016) 89-98. https://doi.org/10.1016/j.electacta.2016.01.112

[31] E. Dannys, T. Green, A. Wettlaufer, C.M.R. Madhurnathakam, A. Elkamel, Wastewater treatment with microbial fuel cells: a design and feasibility study for scale-up in microbreweries, J. Bioproc. Biotech 6:267 (2016) 1-6. doi:10.4172/2155- 9821.1000267

[32] V. Chaturvedi, P. Verma, Microbial fuel cell: a green approach for the utilization of waste for the generation of bioelectricity, Biores. Bioproc. 3:38 (2016) 1-14. https://doi.org/10.1186/s40643-016-0116-6

Chapter 6

Polymeric Membranes and Composites-Innovations, Regulatory Guidelines, Developments for Pollution Control and Environmental Sustainability

V.P.Sharma

CSIR-Indian Institute of Toxicology Research, Lucknow, India

vpsitrc1@rediffmail.com

Abstract

The demand for biocomposites is increasing multifold due to concerns towards environment sustainability and specific features. The basic raw material and additives may remain separate or intermixed through weak bonding. They are generally made up of reinforcing components, fillers, adhesives and matrix which may vary from product to product. The application and demand of composites have been increasing for the fast advancing industries. Innovative and smart composite materials are being designed that have one or more properties that may be significantly changed in a controlled fashion through external stimuli's. Preferences towards utilization of fibers from natural products in composites are also increasing for environmental sustainability. Moreover, the role of nanocomposites and opportunities for composite-metal hybrid materials are critical and needs due care. The functionality with innovation using renewable feedstock are being developed using green polymeric sustainable additives to offer tailor-made, innovative and market-ready polymer solutions. Several natural fiber based composites are rendering help to transform the auto industry by replacing many petroleum-based components. Innovative materials are being developed using functional polymers so that volatile substances may be handled safely. This may help to bind substances and spread without endangering environment or biodiversity. Packaging segment including personal care, pharmaceuticals, food products, water etc. are likely to witness high growth in the future with the lifestyle changes in our society. The continuous monitoring and safety assessment in view of OECD/ ISO norms and specifications, updating as per the National/International guidelines are required periodically to match the state-of-the-art developments.

Keywords

Polymers, Composites, Regulatory, Improvements, Risk Management

Contents

1. Introduction

Globally, developments in the plastic industry have been changing to meet the unique specifications for biomedical and other applications [1-12]. The graphite based nanomaterials, viz., carbon nanotubes, fullerenes or graphene have received interest in the domain of sensing and biomedical therapy. Scientists and engineers have attempted to understand living organisms as they are skillful innovators and fabricators of materials, driven by the forces of evolution [3-21].

The plastic composite packaging renders the advantage of lesser weight and variety of shapes and variance in color or texture. The salient functional features are the strength-to-weight ratio, improved barrier properties and self-sealing. The expectations of consumers are fast changing with lifestyle and architectural expectations of quality products in minimum expenditure for building materials, aerospace and packaging solutions. They may provide specific properties and performance in a wide range of usage applications due to good mechanical properties over conventional materials.

The applications of reinforced plastics in biomedical implants are also requiring physicochemical and biological compatibility characteristics beside no toxicity for long intervals. The method of moulding is improved depending on the item design requirements taking the help of computer designed models. Several advanced products at the modern shelf of materials science need to be functional and prove performance through validated results. These products manufacturers need to demonstrate through experimental data the compliance to regulatory guidelines and produce evidence linked function meeting challenges of safety and economy. The common intentions are to reduce petroleum dependence, biocontent enhancement and commercial production with changing markets demands of composites for varied usages. Synthetic polymers may be developed in the future to recognize selective sites using molecular imprinting processes for environmental control such as nitrate or phosphate in water filtration facility. Now renewable polyols are being synthesized from renewable resources so as to deliver performance at affordable cost. Some are made from recycled polyurethane scrap foam and may function as replacement for petroleum-based products to serve as emulsion formulations, pearlizing agents such as shampoos, shower gel, face cleansers, liquid soap with opacifying appearance in personal wellness industries towards conscientious approaches.

The gels with polymeric base constitution may be cross-linked networks with action like foam or swelling capabilities. The labile characteristics of the solvent enable rapid and reversible swelling or shrinkage in response to a small change in their environment. Polymer structural composites may be engineered materials comprising of high-strength fibers impregnated with a polymer matrix to form a reinforced layer. They may be bonded together with other layers under preset environmental conditions heat and pressure to form an orthotropic laminates. The most common gel forming polymers are polyvinyl alcohol, polyacrylicacid and/ or polyacrylonitrile. Microsized gel fibers may contract in milliseconds, while thick polymer layers may require much longer duration to react. Neural differentiations on graphene have attracted application in stem cell research due to usage of graphene-based neuronal tissue engineering. These aspects may be helpful in the regenerative therapy of various incurable neurological disorders and the fabrication of neuronal networks for therapeutic purposes.

Materials after short lifespan may reveal degradation due to physical, chemical, and/or biological stimuli. Moreover, external static or dynamic forces, internal stress states, corrosion, dissolution, erosion, or biodegradation leading to deterioration of the materials structure. In future years, sophisticated classes of smart materials are expected which may emulate biological systems with capability to select and execute specific functions intelligently and respond to variability in the environment. The burning of composites generates a complex mixture of combustion products comprising of volatile gases, organic vapors and fibers or small particulate matters. The emanation of mixture of toxic gases, particulates and other combustion products of unknown identity may pose unique protection problems due to inadequate knowledge of the hazards related to composite materials. There are concerns related to health complications anticipated through incomplete combustion of flammable products or textiles. There is a vital need to assess the implications of any synergistic interactions between the varied combustion products and other intrinsic factors. Accidental cases or incidences provide experiences to relate wind direction, humidity and metrological conditions of specific locations. It is necessary to scientifically understand the implications of chemical treatment on the viscosity, burning behavior, energy absorption, flammability and biodegradability properties of specific composite materials.

National or International guidelines, viz., BIS, DIN, ASTM's, ISO, EU have composite standards which are effective for the evaluation and determination of the degradation pattern or impact on properties of several forms of composite materials. These composite standards are also helpful in guiding manufacturers and users of such materials in their proper fabrication and testing for the assurance of their quality viz laminates, fiber-reinforced polymer matrix composites, fiber reinforced metal matrix composites, sandwich beams, plates or fabric-reinforced textile composite materials.

The continued efforts in a holistic manner are needed for the development of composites for minimization of environmental impact of polymer composites fabrication. The state of art knowledge and safety assessment is essential for synthesis to final disposal for a sustainable environment for fiber-polymer composites [12-19]. The salient standards related to plastic and polymers based are detailed below in Table 1.

Table 1   Salient standards on plastics and polymeric products.

S.No	Specification/ Guideline Number	Title/ Details
1.	ISO 16620:2015 [Part 1-5]	Biobased Content-General Principles to declaration of biobased carbon content, biobased synthetic polymer content and biobased mass content
2.	ISO 14125:1998 Reaffirmed 2013	Fibre Reinforced Plastic Composites-Determination of flexural properties
3.	ISO 14129:1997 Reaffirmed 2012	Fibre-reinforced plastic composites -determination of the in-plane shear stress/shear strain response, including the in-plane shear modulus and strength, by the plus or minus 45 degree tension test method
4.	ISO 14885:2012	Biodegradability evaluation of Polymers-determination of the ultimate aerobic biodegradability of plastic materials under controlled composting conditions - method by analysis of evolved carbon dioxide- Part 1: General method
5.	ISO 14885:2007 Reaffirmed 2010	Determination of the ultimate aerobic biodegradability of plastic materials under controlled composting conditions- method by analysis of evolved carbon dioxide - Part 2: gravimetric measurement of carbon dioxide evolved in a laboratory-scale test
6.	ISO 15114:2014	Fibre Reinforced Plastic Composites- determination of the mode II fracture resistance for unidirectionally reinforced materials using the calibrated end-loaded split (C-ELS) test and an effective crack length approach
7.	ISO16620:2015	Plastics - biobased content- Part 1: General principles
8.	IS: 10106 [Part 1-4]	Packaging Code
9.	IS: 10171:2013	Guide on suitability of plastic for food packaging
10.	IS 14534:2016	Guidelines for recycling of plastics

Abbreviations ISO: International Organization of Standardization; **IS**: Indian Standards

2. Membranes: natural, biological or synthetic

Membranes may serve as selective barrier for separation purposes and allow selected moieties to pass through but restricts others. These are developed or exists naturally for use in laboratories, industrialists or natural systems. They may have preparation from organic or inorganic materials and have different thickness and homogeneity. The filtration may be aided due to pressure or concentration gradient. Permeate is the liquid that traverses via membrane while feed stream is the term assigned for influent and concentrate/ retentate for the liquid having the retained constituents.

2.1 Selectivity and classification

The degree of selectivity of a membrane is dependent on the size of pore and categorized accordingly. Membranes may be neutral or as charged particle surface with homogenous film, amorphous or heterogeneous solids constitution. It may be facilitated by concentration, chemical or electrical gradients and pressure of the membrane. Membranes may be classified into synthetic or biological membrane. Science and technological innovation are playing decisive roles in improving the quality of the environment and health. The interaction between polymer and solvent may also impart porous structures beside the processing conditions during fabrication. In case of dense membranes viz polytetrafluoroethylene or cellulose esters, a glassy or rubbery surface may exist depending on glass transition temperature. The porous membranes have multiple usages in day to day life ranging from water filtration to dialysis.

2.2.1 Microfiltration (MF)

Microfiltration has the capability to remove particles higher than 0.08-2 μm and operates within a range of 7-100 kPa. Conducive microfiltrations of composites are being prepared with unique properties to control fouling as well as strength to the matrix. The processes are applied for removing residual suspended particles; remove bacteria in order to condition the water for effective disinfection and as a pre-treatment step for reverse osmosis. Relatively recent developments are membrane bioreactors which combine microfiltration and a bioreactor for treatment biologically. Thus sustainable bioreactors are of demand for waste water and effluent treatment for complex process challenges e.g. protein fractionation, microbial removal etc.

2.2.2 Ultrafiltration (UF)

The ultrafiltrations are used by manufacturers or academicians for concentration or purification protein fractions i.e. macromolecular in the range of 10^3 - 10^6 Daltons solutions. These are applied for virus removal due to its high throughput even under ambient or desired environmental conditions. The principal concern regarding the adoption of conventional ultrafiltration membranes for virus removal is the possibility of the virus passing through abnormally large pores or surface imperfections on the membrane surfaces. In addition, they are able to remove viruses and some endotoxins from bioproducts and may be customized for special configurations to attain specific application objectives. Multistage ultrafiltrations are used as per requirement by the stakeholders.

2.2.3 Nanofiltration (NF)

Nanofiltration is in the range of 1-10 nanometers. It is utilized for the removal of selected dissolved constituents from wastewater. They are in great demand due to stringent regulations and to manage costs towards manufacturing and waste discharge management. It is primarily developed as a membrane softening process which offers an alternative to chemical softening. It may be used with objectives to minimize particulate and microbial fouling of the RO membranes by removal of turbidity and bacteria, prevent scaling by removal of hardness ions, lower the operating pressure of the reverse osmosis process by reducing the feed-water total dissolved solids concentration.

2.2.4 Reverse osmosis (RO)

These are commonly used after treatment and filtration for desalination or for the removal of dissolved constituents from wastewater. These may require high pressures to produce deionized water (850-7000 kPa) and considered effective with limitation of wastage of water according to few experts. The salient elements of any membrane processes relates to influence of the parameters such as membrane permeability, operational driving forces per unit area, fouling and cleaning of the membrane surface based on the overall permeate flux.

3.   Green economy and composites

The concepts of green economical concepts are attaining relevance to meet sustainability issues and steps to have minimal toxicity [19]. The nanotechnologists are exploiting characteristics of nanomaterials for green-innovative applications which are energy efficient as well as economically and environmentally sustainable. We look forward for positive impact on manufacturing processes and paradigm shift in the global economy. These solutions may provide better understanding to have opportunities for reducing pressure on raw products handling for generation of renewable energy. This may also contribute to improving delivery systems in a time targeted manner with reliability, efficiency and efficacy in safe mode. We plan to use unconventional water sources or nano-enabled construction products therefore providing better ecosystem and livelihood conditions. The chemistry of membranes is very interesting and offers hydrophilicity, ionic charge, thermal resistance or biocompatibility. Attempts are being made to meet sustainable agricultural production and safe groundwater with secondary effluent. Or through combined two-stage membrane pilot system which is a combination of ultrafiltration and reverse osmosis. Improved agricultural yields may be attained via application of effluent having minimal content of dissolved solids as compared with secondary effluent without any further treatment or high pollution load. We may reduce solvent consumption in the chromatographic analysis of organic and inorganic analytes with increased awareness and improved equipment's using less quantity of sample requirements for analysis.

3.1 Innovative methods for fabrication of composites

Several innovative methods may be developed in view of raw material selection, irrigation techniques, monitoring, and management. There is resurgence of interest in the development of the wood-plastic composites industry and nanocelluloses composites. The composites of the next generation may have less time consuming processes with improved performance and quality. The latest demands and desire to decrease petroleum dependence; increase biocontent, commercial production of nanocelluloses with changing markets will play a major role in the selection of these composites[4-16]. Cellulose nanowhiskers are obtained from plant fibers with properties that are essential towards suitability for their use in the development of bionanocomposites. It has been demonstrated that the capability to enhance the properties of a polymer matrix with low filler loading. Poly (lactic acid) bionanocomposites may be prepared using the solution casting technique, by incorporating the PLA with them or obtained from an oil palm. Fourier transform infrared spectroscopy results revealed no significant changes in the peak positions thus indicating the CNW into the PLA did not result in any significant changes in the chemical structure of the PLA. The composites of ethylene-vinyl acetate (EVA) reinforced with graphene platelets may be fabricated and characterized for parameters viz morphological, thermal, mechanical, electrical properties to moisture absorption of the composites. Tensile strength and elongation of the composites may be improved by graphene platelets.

Growing consumerism and government initiatives for safe packaging models may also contribute in increasing demands for use of less expensive and convenient packaging and pushing the market towards biobased packaging. The stakeholders or customers are increasingly looking to purchase items which are suitable for carrying, long lasting and easy while storing with great versatility. These have been the material of choice for packaging in the manufacturing hub areas and mainly II and III tier cities of the developing countries.

4.   Water and membrane technology for purification - applications

Water is a boon for life and the greater the boon, the greater the conservation is required as it is a precious resource. Without water there could be no development of a civilization and thus sustainable usage is required. There is no denying that rivers are highly revered but it is time to translate our reverence for rivers into actual conservation efforts. In a largely agrarian country rivers have cultural, spiritual and religious connotations. With greater industrialization and urbanization, rivers have become repositories of urban waste and industrial effluents. We need to strengthen coordination between states and stakeholders.

Water quality management is vital and it is difficult to design a strategy which is applicable in all regions due to extreme variations ranging from geographical, chemical, hydrological, political and other man made issues. The long-term development of the global water situation is associated with the growth of the population and global climate change. Water is in acute shortage and thus we need for efficient water sustainable chemical industry with option of green chemical approaches.

The presence of micro-pollutants in wastewater and in drinking water and its sources are posing both environmental and health concerns. The developments of weak polyelectrolyte multilayer based hollow fiber nanofiltration membranes are in progress to remove micro-pollutants from aqueous sources. The charge density of weak polyelectrolytes may be controlled by the pH of the coating solution, providing an additional parameter to tune the performance of the prepared membranes. The entrepreneurs or young professionals may develop solutions for cost effective, better water or effluent treatment with state-of-the-art knowledge of developed nations. Numerous strategies regarding water management are centered towards reducing the consumption, recycling of water and reuse of the precious natural resources. The reuse of water should be encouraged by closer cooperation between the industrial sector, agriculture sector and the local authorities of urban areas. The development of closed loop recycles and reuses processes with modifications in processes or product line of manufacturing may lead to reduction of consumption of water. The advanced know-how of the processes at an industrial site may help in reducing of wastage or leakages of water. The conventional outdated type of pipeline distribution needs to be replaced with new ones. Solutions regarding water quality deal with the continued monitoring and prevention of pollution. Advanced technologies dealing with specific water quality problems have to be developed for the conservation and improvements in water quality.

Wastewater treatment procedures must also focus on the recovery of valuable resources, both organic and inorganic beside effective and reliable remediation steps. The political, legal and environmental driving forces need to be active concerning education and public awareness. Manufacturers may help in creating mass awareness regarding available technologies, reuse of alternative water use, conservation, water harvesting and preventive measures.

Cooperation from other interrelated sectors may help to implement usage of glass fibrer reinforced plastic piping system for irrigation, transportation and distribution to remote places. Education or trainings to employed manpower in the chemical and water sector would be beneficial. The development of advanced membranes with specific functions based on nanotechnological models may help in creation of new engineering concepts.

5. Recent developments in membrane technology, strategies for improving mucosal drug delivery and use of nanotechnology

There are great potential for improving drug delivery devices for varied routes of administration to mucosal surfaces, nanoscale delivery systems seem to be promising tools for further investigations. Mucosal delivery systems establish promising features as they are patient compliant, non-painful and have an accepted administration route.

The Ion Conducting Polymers [ICPs] are fabricated to exhibit the required combination of properties such as high ionic conductivity. The control ICPs may be modified to simultaneously improve their long-term stability and decrease water content, while maintaining the original conductivity. In concert with our fuel cell system consultant, we will assess the feasibility of this technology for use in UAV platforms. The developments of molecularly imprinted polymers are still in the infancy stage. There are potential opportunities for these materials in the drug delivery systems with various administering routes via transdermal, ocular or oral pathways. MIPs are versatile tools in the modern materials science with properties that may be used in construction of the future drug delivery devices by providing some improved delivery profiles, prolonged releasing times and extended residency of the drug. Moreover, MIPs may help to release the drugs in the regulated way, which is extremely advanced and currently required in modern drug delivery systems. Comprehensive analysis of the structure and binding experiments with few structurally related compounds has been demonstrated to show recognition ability and a higher affinity for flurbiprofen in aqueous solutions. They are selective materials with an ability to provide an appropriate enantiomeric form of the drug. Future perspectives for MIPs applications as drug delivery forms are very promising for making it individual specific. They have been used as blood substitutes, tissue regenerators, various auxiliary materials or excipients. Efforts are being taken for developing guided tissue engineering based techniques for controlling periodontal diseases beyond the approaches of traditional concepts. The preventive and predictable reconstruction of periodontal structure, innate organization and function of teeth still continues to be a challenging task for the clinicians. The development of reproducible and clinically safe tissue repair and regenerative models is cumbersome. We require efficient treatment methods and updated public health data for designing strategies to safeguard health and effective public distribution system. The region and country specific therapeutic approaches should be promising to the target groups.

6.   Biofilms or exo polysaccharide (EPS)

Biofilms of high molecular weight which are made of exopolysaccharides may serve as scaffolds. Biomedical device related - matrix enclosed biofilm infections are the new aspects of concern for medical fraternity as the pathogens may be heterogeneously distributed. They are considered to persist and not cleared by host immune defenses or antibiotic therapies. On the other side nanoparticles have opened a new vista of medical applications for biopolymers for development of drug controlled-release systems. The exopolysaccharides have unique and complex chemical structures with appealing physicochemical and rheological properties and novel functionality. The microcrystalline powdered cellulose is used as inactive fillers in drug tablets and as thickeners or stabilizers for processed foods. The future R&D has to be directed towards environmentally friendly designs.

The new chemical identities with tailor made designs are required to reduce environmental footprints. Progresses are desired in the development of biomimetic materials or films with structural configurations for treatment of heavily loaded effluent with variety of toxicants. They may be gel based, oxidant resistant or ceramic membranes with catalytic properties and superior membrane capacity. It is estimated that 700 million people currently live in areas where there is a scarcity of water and this number may multiply by factor of two or three in next few decades and thereafter under conditions of water stress. Safe packaging of potable water is necessary from health point of view and body needs.

7.   Other applications for improved quality of life

Polymeric composite materials are used in single and multi-modal implantable biomedical devices for post-operative treatment of breast cancer. Localized heating using implantable magnetic thermo seeds may be considered one of the latest treatment modalities in terms of fundamental research and clinical research. However we need to assess the drug release characteristics of such nanocomposite system under *in-vitro* and then *in-vivo* conditions based on toxicological parameters of Organization for Economic Cooperation and Developments (OECD) in Good Laboratory Practices (GLP) compliant laboratories[13-15]. It may be facilitated by greater focus on R &D and innovations with due regard to environmental concerns and attention towards harmonious and sustainable approach.

The step wise burning of composites may take in five progressive steps ranging from heating, decomposition, ignition, combustion and propagation. The flame retardants may be used for phenolic, ceramics, intumescents, glass mats, silicone, etc. The coatings and additives may be used with enhanced fire barrier treatments wherein underlying components are protected against flux and heat[8].

Conclusion

Polymers and polymeric items are comparatively economical, strong, and durable with considerable benefits to humanity. They may serve for welfare of modern generation with few limitations. The segregation, transportation and disposal of plastic waste materials have caused adverse implications on water bodies and biodiversity. Improved waste management practices may help in reducing pollution and generation of energy to appropriate levels. The biobased reinforced polymer composites are anticipated to depict beneficial characteristics. It provides advantages for utilization in commercial applications and acceptance by customers at affordable prices.

It is vital to set up an extended environmental and health education program in all communities that will accompany younger generations from the beginning of their education in order to make them conscious players in protecting the environment and their own health. The size of the problems requires that a joint effort be made at an international level in the political realm and in the scientific-cultural realm in order to share experiences and sensitivity and to seek joint solutions. Internationalization produces exchange at various levels, and expands knowledge to a broad audience; as such, it is an instrument for personal gratification, for professional fulfillment and, without being rhetorical, of social progress.

With knowledge of the internationalization of markets and economies it becomes vital to include the global best minds at common platform to see the better world in coming years. The main objective is to enhance excellence in individual countries by expanding and intensifying cooperation. Several impediments to improving the environment and the sustainable exploitation of resources stem from extremely poor knowledge on the part of a very large proportion of the world's population. We should be aware of advantages of extended and effective environmental education for new generations, starting as early as possible to promote and sustain the proper actions, both locally and globally.

There are ever growing needs for R&D to extend the application avenues and push the boundaries of sustainable, compatible, readily available materials, ancillary fixtures and components properties. The objective of the toxicity or biological evaluation of materials, intended for the fabrication of medical devices or scaffolds are to investigate the potential biological hazards by careful observations for unexpected adverse reactions or events in humans during clinical use. The range of biological hazards may be divided into categories viz short term assessment including acute toxicity, irritation, sensitization, haemolysis and thrombogenicity and long-term assessments include subchronic and chronic toxicity, sensitization, genotoxicity, carcinogenicity, effects on reproduction including teratogenicity and tissue material interaction.

Acknowledgement

The author is thankful to the management of CSIR-Indian Institute of Toxicology Research, Lucknow, India.

References

[1]   M. Cole, P. Lindeque, C. Halsband, T.S. Galloway, Microplastics as contaminants in the marine environment: a review, Mar. Pollut. Bull., 62 (2011) 2588-2597. https://doi.org/10.1016/j.marpolbul.2011.09.025

[2]   D. Depan, Biodegradable polymeric nanocomposites- advances in biomedical applications, CRC publication, 2016. ISSN 13-978-1-4822-6051-9.

[3]   FICCI report Proc 2nd, National Conference on Plastic Packaging-The Sustainable Future; A report on Plastic Industry; January (2016).

[4]   S. Ghorai, A. Sinhamahpatra, A. Sarkar, A.B. Panda, S. Pal, Novel biodegradable nanocomposite based on XG-g-PAM/SiO2: application of an efficient adsorbent for Pb2+ ions from aqueous solution, Bioresour. Technol., 119 (2012) 181-190. https://doi.org/10.1016/j.biortech.2012.05.063

[5]   H.J. Endres Bioplastics, Adv. Biochem. Eng. Biotechnol., 2017 Apr 4.

[6]   J. Hammer, M.H.S. Kraak, J.R. Parsons, Plastics in the marine environment: the dark side of a modern gift, Rev. Environ. Contam. Toxicol. 220 (2012) 1-44. https://doi.org/10.1007/978-1-4614-3414-6_1

[7]   H.M. Koch, A.M. Calafat, Human body burdens of chemicals used in plastic manufacture, Phil. Trans. R Soc. B, 364 (2009) 2063-2078.

[8]   K. Kan-Dapaah, N. Rahbar N, W. Soboyejo; Polymeric composite devices for localized treatment of early-stage breast cancer; PLoSOne, 12 (2017) 1-11. https://doi.org/10.1371/journal.pone.0172542

[9]   K. Kan-Dapaah, N. Rahbar, A. Tahlil A, D. Crosson, N. Yao, W. Soboyejo W; Mechanical and hyperthermic properties of magnetic nanocomposites for biomedical applications, Mech. Behav. Biomed. Mater. 49 (2015) 118-128. https://doi.org/10.1016/j.jmbbm.2015.04.023

[10]   Layth Mohammed, M. N. M. Ansari Grace Pua, Mohammad Jawaid, and M. Saiful Islam A Review on Natural Fiber Reinforced Polymer Composite and Its Applications; International Journal of Polymer Science; Article ID 243947, 2015 (2015) 1-15.

[11]   M. Marcos-García, P. García-Fraile, A. Filipová, E. Menéndez, P.F. Mateos, E. Velázquez, T. Cajthaml, R. Rivas, Mesorhizobium bacterial strains isolated from the legume Lotus corniculatus are an alternative source for the production of Polyhydoxyalkanoates to obtain bioplastics. Environ Sci Pollut Res Int., (2017) Jun 7.

[12]   M. Scavone,, I. Armentano , E. Fortunati , F. Cristofaro, S. Mattioli, L. Torre, J. M. Kenny, M. Imbriani, C.R. Arciola, L. Visai, Antimicrobial properties and cyto-compatibility of PLGA/Ag nanocomposites materials, 9 (2016) 1-15.

[13]   C.J. Moore, Synthetic polymers in the marine environment: a rapidly increasing, long-term threat, Environ. Res. 108 (2008) 131-139. https://doi.org/10.1016/j.envres.2008.07.025

[14]   N. Uddin, Ed., Developments in Fiber-Reinforced Polymer Composites for Civil Engineering, Elsevier, 2013, ISBN: 9780857092342. https://doi.org/10.1533/9780857098955

[15]   Naoji Matsuhisa, Daishi Inoue, Peter Zalar, Hanbit Jin,Yorishige Matsuba, Akira Itoh, Tomoyuki Yokota, Daisuke Hashizume and Takao Someya, Printable elastic conductors by in situ formation of silver nanoparticles from silver flakes, Nature Materials, 2017. https://doi.org/10.1038/nmat4904

[16]   M.S. Muhamad, M.R.,Salim, W.J.,Lau, Z.Yusop, A review on bisphenol A occurrences, health effects and treatment process via membrane technology for

drinking water (2016), Environ. Sci. Pollut. Res., 23 (2016) 11549-11567. https://doi.org/10.1007/s11356-016-6357-2

[17]   P. Pavani and T Raja Rajeswari Impact of Plastics on Environmental Pollution, Proc. "National Seminar on Impact of Toxic Metals, Minerals and Solvents leading to environmental Pollution - 2014" J. Chem. Pharm. Sci. 3 (2014) 87-93.

[18]   T. Zhang, W. Wang, D. Zhang, X. Zhang, Y. Ma, Y. Zhou, L.Qi, Biotemplated synthesis of gold nanoparticles-bacteria cellulose nanofibers nanocomposites and their application in biosensing, Adv. Funct. Mater., 20 (2010)1152-1160. https://doi.org/10.1002/adfm.200902104

[19]   Urbini G, Internationalization, education and technological innovation: three key factors to improve the quality of the environment and public health, Rev. Ambient. Água, 10 (2015) 1-8.

[20]   V.A Cataldo, G. Cavallaro, G. Lazzara, S. Milito, F Parisi, Coffee grounds as filler for pectin: Green Composites with Competitive performances dependent on the UV irradiation, Carbohyd. Polym. 170 (2017) 198-205. https://doi.org/10.1016/j.carbpol.2017.04.092

[21]   Z. Ding, C. Yuan, X. Peng, T. Wang, H.J. Qi, M.L. Dunn; Direct 4D printing via active composite materials, Sci Advances, 3 (2017) 1-6. https://doi.org/10.1126/sciadv.1602890

Chapter 7

Exploring a Green Bio-Nanocomposite for the Removal of Crystal Violet Dye

Anam Mirza, Rais Ahmad[*]

Environmental Research Laboratory, Department of Applied Chemistry, Aligarh Muslim University, Aligarh, 202002, India

* rais45@rediffmail.com

Abstract

In the present chapter alginate/activated carbon bio-nanocomposite was evaluated for the removal of crystal violet dye from aqueous solution. The bio-nanocomposite was characterized by SEM, EDX and TEM analysis. In batch adsorption experiments effects of various parameters such as pH, contact time, concentration and temperature were performed in order to evaluate the adsorption process. The adsorption was best described at pH 7.2, contact time 300 min and temperature 293 K. The experimental data were evaluated for isotherm and kinetic studies. Langmuir and Pseudo-second order models were found to be the best-fitted model. Therefore, the present bio-nanocomposite was found to be an excellent adsorbent for the removal of toxic dye from wastewater.

Keywords

Adsorption, Removal, Dye, Bionanocomposite, Adsorbent

Contents

1.   Introduction

Due to rapid industrialization and globalization, water contamination has been a major area of concern for environmental researcher [1-3]. The effluents from different industries like textile, food processing, cosmetics, rubber, plastics, paper and pulp, tanneries and paint industries are the major sources of water pollution containing dyes [1,4]. Dyes are one of the most notorious organic contaminants. They have very complex structure and are aromatic in nature. Thus, they are mutagenic and carcinogenic to mankind [4-6]. The dye also, has adverse effects on the water bodies as the color interfere with the penetration of light into the water [4]. Hence, to defend ecological, biological and industrial environment on a global scale, dyes must be efficiently removed from the discharged wastewater.

Crystal violet is a tri-phenylmethane dye [4]. Exposure to the dye causes eye irritation and can cause permanent injury to the cornea and conjunctiva. It can also cause skin and digestive tract irritation. In extreme cases, it may lead to respiratory and kidney failure and might cause permanent blindness [7]. Therefore, keeping in view the hazardous effects of crystal violet dye, it has been selected as a potential dye for necessary removal.

Various technologies like physical, chemical and biological methods have been reported for the removal of dyes from aqueous solution. Most of the methods are unsuitable and inefficient under certain conditions. Adsorption method is the most preferred method for the removal of dyes from aqueous solution due to its cost effective nature, easy operation, insensitivity to toxic substances, ability to treat concentrated forms of the dyes and the possibility of reusing the spent adsorbent [8-12]. Various studies have already been reported that includes the low-cost adsorbent but among those, activated carbon is an effective adsorbent for the removal of dyes from aqueous solution due to its high adsorption capacity [13]. Herein, we have reported the use of *Tea activated carbon along with alginate*. The use of environmental friendly adsorbents with higher adsorption capacity has attracted extensive interest to researchers all over the world. Nanocomposites containing biopolymers have received great attention due to their non-toxic, biodegradable and biocompatible nature and are applied in the treatment of water pollution from toxic and carcinogenic dyes [14]. Of particular concern in biopolymers is the sodium alginate, which has its unique colloidal properties, excellent hydrophilicity, binding ability, low-cost, biocompatibility and renewability. Sodium alginate is a salt of alginic acid, which consists of guluronic and mannuronic acid residues [8, 9].

The present chapter aims to synthesize a novel and eco-friendly *Alginate/Activated carbon bio-nanocomposite* and evaluates its adsorption capacity for the removal of crystal violet dye by varying the experimental factors such as (pH of the solution, contact time, the initial concentration of dye and temperature) that are responsible for the adsorption process. The synthesized bio-nanocomposite was characterized by SEM, EDX and TEM analysis. Isotherm and kinetic studies have also been included.

2. Experimental

2.1 Chemicals

Sodium alginate and crystal violet were purchased from CDH, New Delhi. Raw Tea sample was purchased from the local market. NaOH and HCl were analytical grade. The stock solution of crystal violet dye (1000 mgL^{-1}) were prepared by dissolving the appropriate amount of dye in double distilled water.

2.2 Synthesis of alginate/activated carbon bio-nanocomposite

The tea sample was placed in a silica crucible and kept in a muffle furnace at 600 °C for 4 h. It was then cooled in a desiccator. 2% (w/v) alginate solution was added to the 2 g of activated carbon and the sample was left for vigorous stirring for 24 h at 273 K. The precipitate was then filtered, washed with double distilled water and dried in an oven at 333 K for 3 h. Finally, it was powdered in mortar for subsequent studies.

2.3 Characterization

Surface morphology of the bio-nanocomposite was studied through a scanning electron microscopy (SEM, JSM 6510LV, JEOL, Japan). The elemental characteristics of the bio-nanocomposite were obtained using an Energy-dispersive X-ray (EDX, JSM 6510LV,

JEOL, Japan). The size of the bio-nanocomposite was analyzed using a transmission electron microscopy (TEM, JEM 2100, JEOL, Japan). The absorbance of the crystal violet solution was measured using an UV-Vis spectrophotometer (T70 UV/Vis spectrometer PG instruments Ltd, UK) (λ_{max} = 582 nm). The pH was measured using a pH meter (Perkin Elmer, USA).

2.4 Adsorption studies

Batch adsorption experiments for the adsorption of crystal violet dye using the Alginate/Activated carbon bio-nanocomposite was carried out in a series of flasks with variable conditions depending upon the operating factor. The effect of pH was analyzed by treating 0.01 g of the bio-nanocomposite for 24 h at 273 ± 3K by varying the pH from 1 to 8 with concentration 20 mgL^{-1} 0.5 M. The effect of contact time was analyzed by treating 0.01 g of the bio-nanocomposite at 273 ± 3K by varying the contact time from 5 to 360 min with concentration 20 mgL^{-1} and pH 7.2. The effect of initial dye concentration was analyzed by treating 0.01 g of the bio-nanocomposite at 273 ± 3K by varying the concentration from 5 to 50 mgL^{-1} with contact time 300 min and pH 7.2. The effect of temperature was analyzed by treating 0.01 g of the bio-nanocomposite by varying the temperature from 273 to 293 K with contact time 300 min, concentration 20 mgL^{-1} and pH 7.2. At the end of the experiments, the resulting samples were filtered using Whatman filter paper 1 and the absorbance of samples was then analyzed using a UV-Vis spectrophotometer.

The amount of crystal violet adsorbed onto the bio-nanocomposite was evaluated using the following equations:

The adsorption Efficiency (%), $\% = \dfrac{C_0 - C_e}{C_0} \text{X}100$ (1)

The adsorption capacity (q$_e$), $q_e = \dfrac{C_0 - C_e}{m} \text{X}V$ (2)

where, C_0 and C_e were the concentrations of crystal violet at the initial and time t, respectively; V and m were the solution volume and the amount of crystal violet adsorbed on the bio-nanocomposite, respectively [9, 15].

3.   Results and discussion

3.1   Characterization

SEM micrograph of bio-nanocomposite was depicted in Fig. 1. The micrograph showed highly porous and irregular surface of bio-nanocomposite. The highly porous nature of bio-nanocomposite was responsible for excellent adsorption of crystal violet dye.

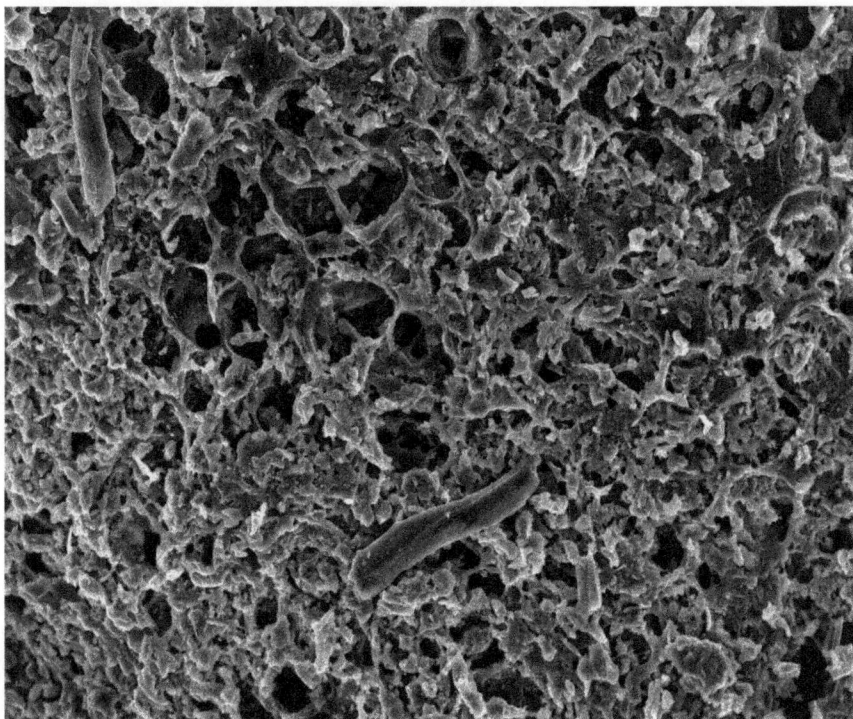

Fig. 1 SEM micrograph of the bio-nanocomposite.

EDX spectra of bionanocomposite was shown in Fig. 2. EDX analysis showed that C and O were the major constituents present in the bionanocomposite. This also implies that the alginate framework of C and O have been successfully modified to the activated carbon framework. The other constituents Mg, Si, K and P were also present that also enhanced the adsorption process.

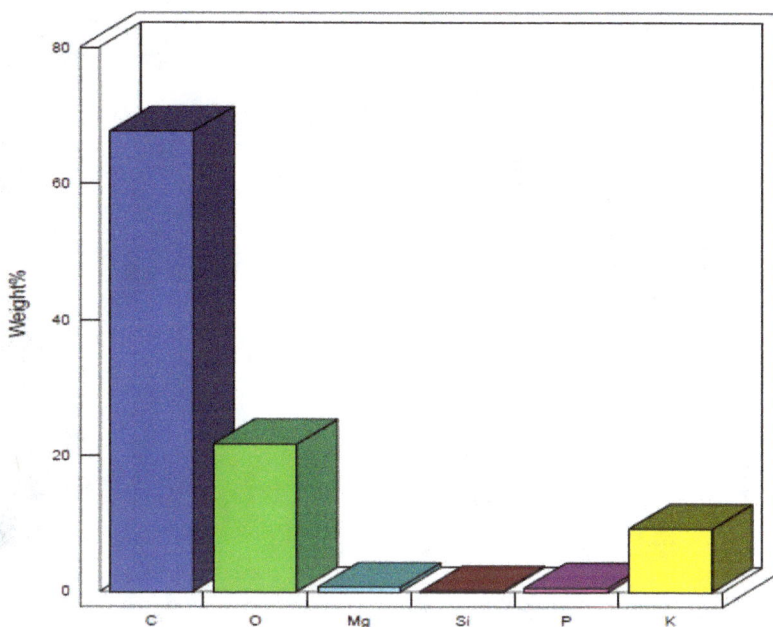

Fig. 2 EDX spectra of the bio-nanocomposite.

The TEM image of the bio-nanocomposite was represented in Fig. 3. The TEM showed the size of the bio-nanocomposite particle in the range of 3-46 nm.

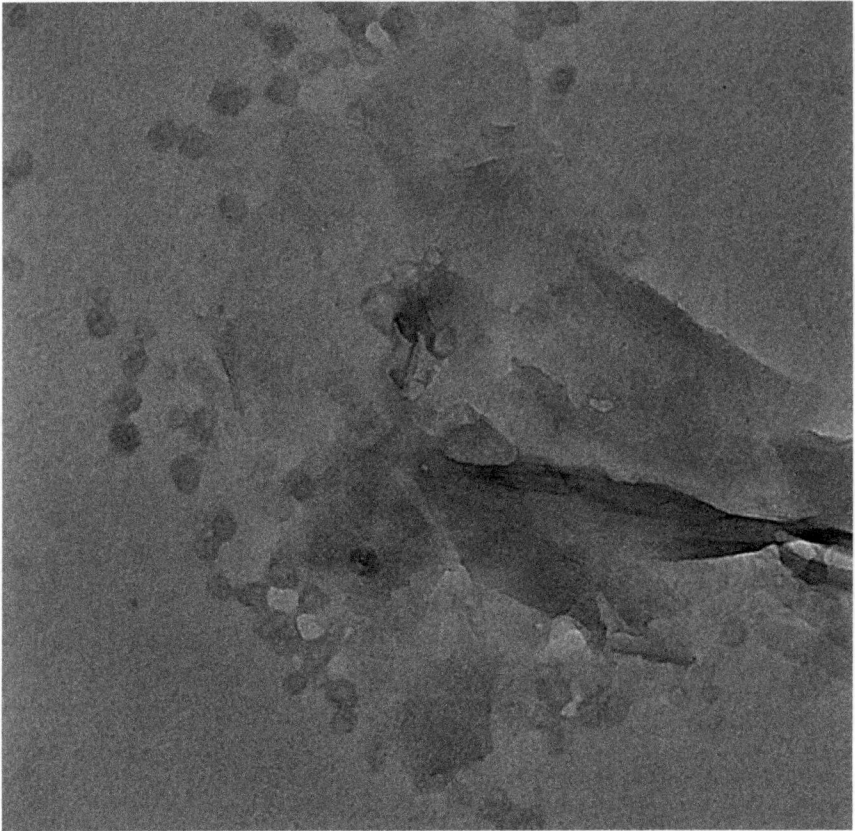

Fig. 3 TEM micrograph of bionanocomposite.

3.2   Effect of pH

The effect of pH on the adsorption of crystal violet by the bio-nanocomposite was represented in Fig. 4. The pH is an important factor governing the adsorption process. It affects the surface charge of the bio-nanocomposite and the degree of ionization of crystal violet present in the solution. Fig. 4 showed that with an increase in pH from 1 to 7.2, the q_e (mgg^{-1}) increases and reaches a maximum at pH 7.2 and then decreases. The results could be explained as- on increasing pH, the functional groups on the surface of the bio-nanocomposite starts deprotonating and thus, there is an electrostatic attraction between the negatively charged surface of the bio-nanocomposite and positively charged dye molecules [1,6].

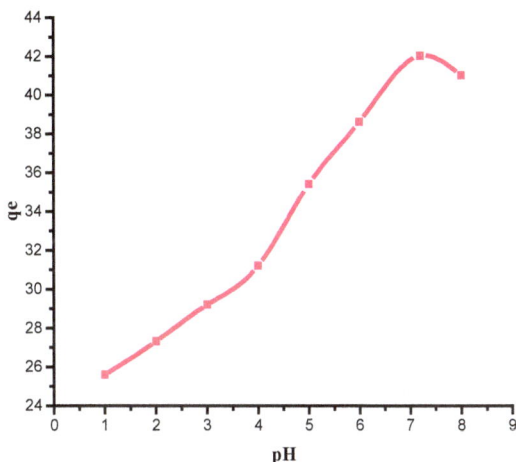

Fig. 4 Effect of pH on the adsorption of crystal violet using the bio-nanocomposite.

3.3 Effect of contact time and kinetic studies

With increase in contact time from 5 to 360 min, the q_t (mgg^{-1}) increases and reaches maximum in 300 min (q_e exp: 52.1 mgg^{-1}) and then equilibrium has been reached. The gradual increase in adsorption capacity with contact time can be attributed to the availability of more adsorption sites that gradually become saturated with time (Fig.5).

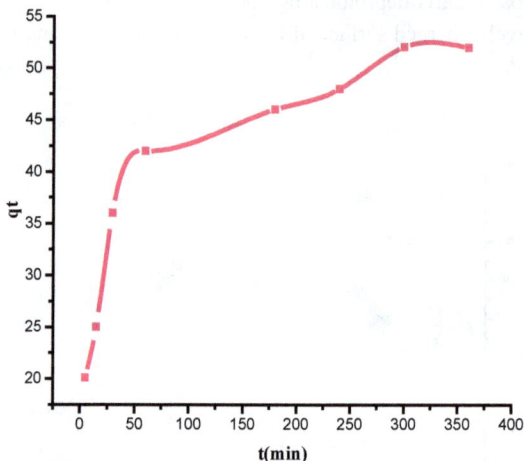

Fig. 5   Effect of contact time on the adsorption of crystal violet using the bio-nanocomposite.

Kinetics of the adsorption process was studied by applying Pseudo-first order and Pseudo-second order kinetic models. The Pseudo-first order kinetic order model can be represented as (Eq. 3):

$$q_t = q_e(1 - e^{-k_1 t})$$
(3)

The Pseudo-second order kinetic model can be represented as (Eq. 4):

$$q_t = \frac{k_2 q_e^2 t}{1 + k_2 q_e t}$$
(4)

where, q_e and q_t were the amount of dye adsorbed at equilibrium and time t (mgg^{-1}), k_1 and k_2 were the pseudo-first (min^{-1}) and pseudo-second order (gmg^{-1}min^{-1}) rate constant, respectively [16].

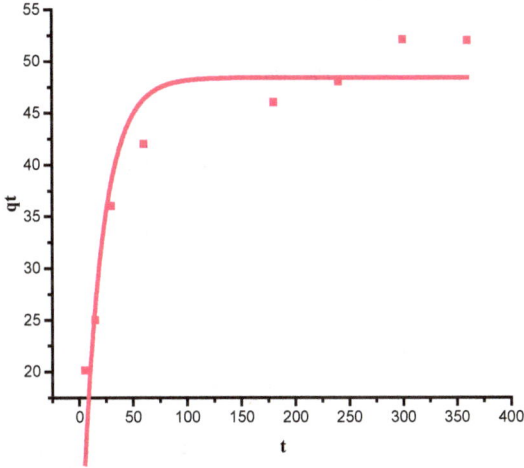

Fig. 6 The pseudo-first order kinetic model for the adsorption of crystal violet using the bio-nanocomposite.

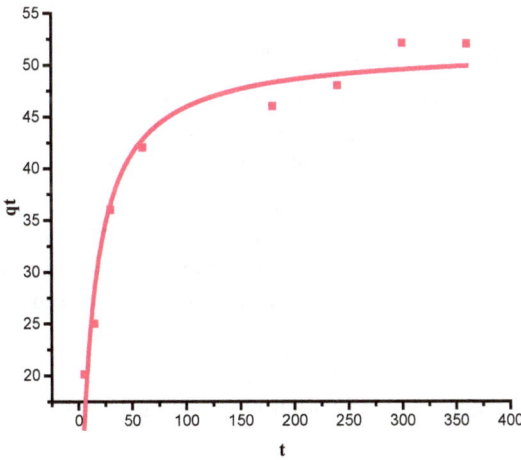

Fig. 7 The pseudo-second order kinetic model for the adsorption of crystal violet using the bio-nanocomposite.

The values of pseudo-first and pseudo-second order kinetic parameters for the adsorption of crystal violet are tabulated in Table 1.

Table 1. Kinetic parameters for the adsorption of crystal violet using the bio-nanocomposite.

Pseudo-first order		Pseudo-second order	
$q_e(mgg^{-1})$	48.383	$q_e(mgg^{-1})$	51.577
$k_1(min^{-1})$	0.052	$k_2(gmg^{-1}min^{-1})$	0.001
R^2	0.843	R^2	0.936

The value of correlation coefficient (R^2) for the pseudo-second order kinetic model was higher than for the pseudo-first order kinetic model. Moreover, the adsorption capacity (cal) obtained from the pseudo-second kinetic model was closer to adsorption capacity (exp). Thus, it can be concluded that the kinetic studies were best fitted by the pseudo-second order kinetic model and the rate-determining step for the adsorption of crystal violet using the bio-nanocomposite was controlled by chemisorption [16].

3.4   Effect of concentration and isotherm studies

With the increase in the concentration of crystal violet from 5 to 50 mgL^{-1}, the adsorption capacity increases from 7.6 to 70 mgg^{-1}. This might be due to the increased concentration gradient between the bulk solution and the bio-nanocomposite.

To examine the relationship between adsorption and aqueous concentration, the experimental data were analyzed for the isotherm studies. The adsorption process was analyzed for two isotherm models; namely, Langmuir and Freundlich isotherm models. The non-linear form for Langmuir isotherm model can be represented as (Eq. 5):

$$q_e = \frac{q_m bC_e}{1 + bC_e} \tag{5}$$

The non-linear form for the Freundlich isotherm model can be represented as (Eq. 6):

$$q_e = K_F C_e^{1/n} \tag{6}$$

where, q_m (mgg^{-1}) and b (Lmg^{-1}) were the monolayer adsorption capacity and Langmuir constant, K_F (mgg^{-1})(Lmg^{-1})$^{1/n}$ and n were related to adsorption capacity and Freundlich constant, respectively [16, 17].

The values of isotherm parameters were tabulated in Table 2.

Table 2. Isotherm parameters for the adsorption of crystal violet using the bio-nanocomposite.

Langmuir isotherm		Freundlich isotherm	
q_m (mgg^{-1})	143.713	K_F (mgg^{-1})(Lmg^{-1})$^{1/n}$	1.640
b (Lmg^{-1})	0.070	n	1.442
R^2	0.963	R^2	0.925

From Table 2, it was concluded that the experimental data were best fitted by the Langmuir isotherm model, in accordance with high correlation coefficient (R^2). This affirmed that there is monolayer adsorption of crystal violet on the homogenous surface of the bio-nanocomposite.

3.5   Effect of temperature

With the increase in temperature from 273 to 293 K, the adsorption capacity increases from 46 to 52 mgg^{-1}. This also showed that the adsorption process was endothermic in nature.

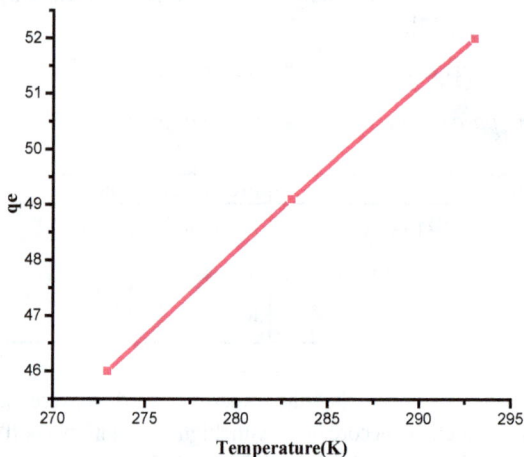

Fig. 8   Effect of temperature on the adsorption of crystal violet using the bio-nanocomposite.

4.   Conclusion

In the present chapter, Alginate/Activated carbon bio-nanocomposite was successfully synthesized and was characterized using SEM, EDX and TEM analysis. The bio-nanocomposite has been excellently applied for the removal of crystal violet from aqueous solution. The maximum adsorption of crystal violet onto the bio-nanocomposite was observed at pH 7.2, contact time 300 min, concentration 50 mgL^{-1} and temperature 293K, respectively. The equilibrium data were analyzed by the Langmuir and Freundlich isotherm models that showed a better fit with the Langmuir isotherm model, implying monolayer adsorption of crystal violet onto the bio-nanocomposite. The batch adsorption kinetics was tested and the pseudo-second order kinetic model was the best-suited model, implying chemisorption as the rate determining step in the adsorption process. The temperature studies showed that adsorption was endothermic in nature. Therefore, it was concluded that the present bio-nanocomposite has been successfully utilized for the adsorption of crystal violet from aqueous solution.

References

[1] H.J. Kumari, P. Krishnamoorthy, T.K. Arumugam, S. Radhakrishnan, D. Vasudevan, An efficient removal of crystal violet dye from waste water by adsorption onto TLAC/Chitosan composite: A novel low cost adsorbent, Int. J. Biol. Macromol. 96 (2017) 324-333. https://doi.org/10.1016/j.ijbiomac.2016.11.077

[2] Qamruzzaman, A. Nasar, Degradation of acephate by colloidal manganese dioxide in the absence and presence of surfactants, Desal. Water Treatm. 55 (2015) 2155-2164. https://doi.org/10.1080/19443994.2014.937752

[3] Qamruzzaman, A. Nasar, Degradation of tricyclazole by colloidal manganese dioxide in the absence and presence of surfactants, J. Ind. Eng. Chem. 20 (2014) 897-902. https://doi.org/10.1016/j.jiec.2013.06.020

[4] G.K. Sarma, S.S. Gupta, K.G. Bhattacharyya, Adsorption of crystal violet on raw and acid-treated montmorillonite, k10, in aqueous suspension, J. Environm. Managem. 171 (2016) 1-10. https://doi.org/10.1016/j.jenvman.2016.01.038

[5] A. Nasar, S. Shakoor, Remediation of dyes from industrial wastewater using low-cost adsorbents. In "Applications of Adsorption and Ion Exchange Chromatography in Waste Water Treatment" Inamuddin, Al-Ahmed A (eds), Materials Research Foundations, Vol. 15 (2017) 1-33. http://dx.doi.org/10.21741/9781945291333-1

[6] R. Ahmad, A. Mirza, Green synthesis of Xanthan gum/Methionine-bentonite nanocomposite for sequestering toxic anionic dye, Surf. Interf. 8 (2017) 65-72. https://doi.org/10.1016/j.surfin.2017.05.001

[7] M. Gholami, M.T. Vardini, G.R. Mahdavinia, Investigation of the effect of magnetic particles on the crystal violet adsorption onto a novel nanocomposite based on k-carrageenan-g-poly(methacrylic acid), Carbohydr. Ploym. 136 (2016) 772-781. https://doi.org/10.1016/j.carbpol.2015.09.044

[8] S. Gokila, T. Gomathi, P.N. Sudha, S. Anil, Removal of the heavy metal ion chromium (VI) using chitosan and alginate nanocomposites, Int. J. Biol. Macromol. (2017), http://dx.doi.org/10.1016/j.ijbiomac.2017.05.117. https://doi.org/10.1016/j.ijbiomac.2017.05.117

[9] R. Ahmad, A. Mirza, Sequestration of heavy metal ions by Methionine modified bentonite/ Alginate (Meth-bent/Alg): A bionanocomposite, Groundwater Sustain. Developm. 1 (2015) 50-58. https://doi.org/10.1016/j.gsd.2015.11.003

[10] R. Ahmad, A. Mirza, Adsorption of Pb (II) and Cu (II) by Alginate-Au-Mica bionanocomposite: Kinetic, isotherm and thermodynamic studies, Process Safety Environm. Protect. 109 (2017) 1-10. https://doi.org/10.1016/j.psep.2017.03.020

[11] S. Shakoor, A. Nasar, Removal of methylene blue dye from artificially contaminated water using citrus limetta peel waste as a very low cost adsorbent, J. Taiwan Inst. Chem. Eng. 66 (2016), 154–163. https://doi.org/10.1016/j.jtice.2016.06.009

[12] S. Shakoor, A. Nasar, Adsorptive treatment of hazardous methylene blue dye from artificially contaminated water using cucumis sativus peel waste as a low-cost adsorbent, Groundwater Sustain. Develop., 5 (2017) 152–159. https://doi.org/10.1016/j.gsd.2017.06.005

[13] E. Asuquo, A. Martin, P. Nzerem, F. Siperstein, X. Fan, Adsorption of Cd (II) and Pb(II) ions from aqueous solutions using mesoporous activated carbon adsorbent: Equilibrium, kinetics and characterization studies, J. Environm. Chem. Eng. 5 (2017) 679-698. https://doi.org/10.1016/j.jece.2016.12.043

[14] Y. Niu, K. Li, D. Ying, Y. Wang, J. Jia, Novel recyclable adsorbent for the removal of copper(II) and lead(II) from aqueous solution, Biores. Technol. 229 (2017) 63-68. https://doi.org/10.1016/j.biortech.2017.01.007

[15] R. Ahmad, A. Mirza, Heavy metal remediation by Dextrin-oxalic acid/cetyltrimethyl ammonium bromide (CTAB)-Montmorillonite (MMT) nanocomposite, Groundwater Sustain. Developm. 4 (2017) 57-65. https://doi.org/10.1016/j.gsd.2017.01.001

[16] R. Ahmad, A. Mirza, Inulin-folic acid/bentonite: A novel nanocomposite for confiscation of Cu (II) from synthetic and industrial wastewater, J. Mol. Liquids 241 (2017) 489-499. https://doi.org/10.1016/j.molliq.2017.05.125

[17] H. Mittal, S.B. Mishra, Gum ghatti and Fe_3O_4 magnetic nanoparticles based nanocomposites for the effective adsorption of rhodamine B, Carbohydr. Polym. 101 (2014) 1255-1264. https://doi.org/10.1016/j.carbpol.2013.09.045

Chapter 8

Poly(N-vinyl-2-pyrrolidone) Stabilized Nanoclusters as Highly Efficient and Reusable Catalyst for the Dehydrogenation of Dimethly Ammonia–Borane

Betül Sen[1], Özge Paralı[1], Süleyman Akocak[1,2], Senem Karahan[1,3], Fatih Sen[1*]

[1] Sen Research Group, Biochemistry Department, Faculty of Arts and Science, Dumlupınar University, Evliya Çelebi Campus, 43100 Kütahya, Turkey

[2] Department of Pharmaceutical Chemistry, Faculty of Pharmacy, Adıyaman University, Adıyaman, Turkey

[3] Department of Chemistry, Faculty of Science, Dokuz Eylül University, İzmir, Turkey

[*] fatih.sen@dpu.edu.tr

Abstract

Addressed herein, we report a highly efficient and facile synthesis of palladium, nickel nanoparticles supported on poly(N-vinyl-2-pyrrolidone) (PdNi@PVP NPs) for the dehydrogenation of DMAB. PdNi@PVP nanoclusters have been synthesized from the reduction of precursors of metals (Pd and Ni) by the microwave assistance method at room temperature with an average particle size of 3.05 ± 0.38 nm. This newly produced monodisperse PdNi@PVP nanoclusters exhibits high durability, reusability, and catalytic performance even after the fourth cycle of dehydrogenation of DMAB reactions. On the other hand, the structure morphology and properties of the nanoclusters were characterized using different analytical methods such as UV-Vis, XPS, XRD, TEM, and the HR-TEM techniques. Besides, the PdNi@PVP NPs catalyst showed good catalytic effectiveness with a high turnover frequency of 561.0 h^{-1} and low Ea value of 37.11 ± 2 kJ mol^{-1} for DMAB dehydrocoupling in ambient conditions.

Keywords

Dehydrocoupling, Facile Synthesis, Microwave, Monodispersity, Nanocatalyst

Contents

1.   Introduction

Nowadays, the trend in science is to obtain new metallic nano-catalysts with different supporting materials. For this purpose, scientists have focused on carbon materials (MWCNTs, rGO, GO, Vulcan Carbon, Activated Carbon etc) because of their significant properties such as chemically stability, electrically and thermally conductivity they can be used in various applications [1-8]. However, some of those are not always perfect because under normal conditions pristine ones are inactive for such an application and not able to dissolve in some solvents, and thus it requires overcoming these problematic issues to make the carbon materials perfect. To overcome these types of problems, two strategies have been used; the first one is the modification of carbon materials and second one is the usage of different polymeric materials. Due to the large surface area of polymeric materials, they can be thought as important materials hence they can be employed in lots of catalytic procedures. One of them is the efficient storage of hydrogen. For this purpose, a variety of materials like amine-boren derivatives are tried for chemical hydrogen storage. More importantly, recent studies have suggested amine-boranes adducts as alternative hydrogen energy sources due to their high hydrogen content [9-47]. Of particular interest, catalytic dehydrogenation of dimethylamine-borane (DMAB) potentially releases 3.5 % H_2 by weight. Furthermore, it is very easy to generate hydrogen by using DMABs at room temperature if there is an appropriate catalyst. A quick literature survey shows that various homogeneous and heterogeneous catalysts have been tested for DMAB dehydrocoupling [12, 19, 28, 32-47]. The aforementioned studies clearly demonstrate that the majority of heterogeneous catalysts with high performance in DMAB dehydrogenation are transition metal NP based catalytic systems. It is well-known that the use of transition metal NPs increases catalytic activity as the fraction of surface atoms increases with decreasing particle size. For this reason, addressed herein, the bimetallic Pd-Ni supported on poly-N-vinyl-2-pyrrolidone (PVP) has been developed as an efficient catalyst for dehydrogenation of DMAB reactions under ambient conditions as a catalyst. The nanocatalyst was prepared by a special method in which both Pd and Ni metals are co-reduced by the microwave assistance method. On the other hand, the characterization of this uniformly dispersed nano clusters were performed by several analytical methods such as UV–Vis, XRD, TEM HR-TEM and XPS analyses.

2.   Experimental methods

2.1   Materials and methods

Ethylene glycol, C_2H_5OH, dimethylamine-borane, K_2PdCl_4, $NiCl_2$, and poly(N-vinyl-2-pyrrolidone) (PVP) have been supplied by Sigma. All glass pieces and other lab materials were washed with large amount of distilled water and then they were cleaned with acetone and then dried. TEM analysis was carried out by a JEOL 200 kV. A specs spectrometer was employed for X-ray Photoelectron Spectroscopy (XPS) for analysis with an X-ray source Kα lines of Mg (1253.6 eV, 10 mA). All XPS peaks were fitted using a Gaussian function and the relative intensity of the species was calculated with the help of the integration of each peak, after smoothing amd subtraction of the Shirley-shaped background. In the XPS spectrum, accurate binding energies (±0.3 eV) have been determined by referencing to the C 1s peak at 284.6 eV. XRD analysis were performed with a Panalytical Empyrean diffractometer with Ultima+theta–theta high resolution goniometer, having an X-ray generator (Cu K∞ radiation, k = 1.54056 Å) and operating condition of 40 kV and 40 mA. UV–Vis analyses were taken by Perkin Elmer Lambda 750. 200–900 nm was selected to gather the data and 1 cm-long a quartz cell was employed. ^{11}B NMR spectra were recorded on a Bruker Avance DPX 400 MHz spectrometer (128.2 MHz for ^{11}B NMR).

2.2   The synthesis of Pd-Ni nanomaterials stabilized by PVP

The Pd-Ni@PVP (poly(N-vinyl-2-pyrrolidone)) nanoclusters were prepared by using the microwave assistance method by using an ethylene glycolic (EG) solution of K_2PdCl_4 and $NiCl_2$. At the beginning, before addition of PVP (2.5 mmol), $NiCl_2$ (0.25 mmol) and K_2PdCl_4 (0.25 mmol) were stirred in ethylene glycol (EG) and PVP mixture, and then a stable suspension was obtained via strong stirring. Ph of this solution was set to 12 by the NaOH–EG solution and then kept in the central point of a microwave oven (1200 W) for 60 s where PVP-EG acted as the reducing agent for $PdCl_2$ and $NiCl_2$ reduction. After that, the resulting mixture was refluxed at 90°C for 2 h. Then, it was cooled to room temperature and the color of palladium-nickel nanomaterials dispersed on PVP was observed as brownish black with having good stability and homogeneity. This color shows that Pd^{2+} and Ni^{2+} ions were reduced to Pd(0) and Ni(0) to form bimetallic nanoparticles of the same. Analytical investigations of the prepared Pd-Ni@PVP NPs showed their reusability, their catalytic performance, stability and efficiency for DMAB dehydrogenation and, investigation of E_a of Pd-Ni@PVP NPs have been analyzed in detail (see the supporting information for all detail).

2.3   The catalytic performance of Pd-Ni@PVP for DMAB dehydrogenation

The dehydrocoupling of DMAB was performed in a typical jacketed reaction flask connected to the water-filled cylinder glass tube under dry nitrogen atmosphere. The jacketed reaction flask was vacuumed for at least 15 min and filled with nitrogen to remove any trace of oxygen and water present prior to all the catalytic reactions. Then, all reactions were started by closing the flask thermostated at 25.0 ± 0.1 °C and turning on the stirrer at 1000 rpm simultaneously. The catalytic activity of Pd-Ni@AC NPs in the solvent-free dehydrogenation of DMAB was determined by measuring the rate of hydrogen generation. DMAB (2 mmol) was added on Pd-Ni@AC NPs (25 mg) in a jacketed reaction flask under vigorous stirring for 5 h at 25.0 ± 0.1 °C. Hydrogen gas generation from the catalytic solid reaction mixture was tracked by using a typical water-filled gas burette system that shows the displacement of water level in the gas burette for every 5 min until no more hydrogen evolution ae observed. The experiment was stopped when H_2 generation ceased, and the reactor was disconnected from the water-filled tube and the hydrogen pressure was released. For the reusability experiments of Pd-Ni@AC, 2.0 mmol DMAB at 25.0 ± 0.1 °C was added into a new Schlenk tube which included Pd-Ni@AC, which was sealed under a dry nitrogen atmosphere at the end of the dehydrogenation reaction. The solid mixture was precipitated with cold hexane (10 mL; added under N_2 atmosphere) and the supernatant solution was removed by filtration. The solid was further washed with hexane (3×20 mL) and dried under vacuum, giving the isolated colloid as a dark brown powder.

3.   Results and discussion

The initial characterizations of highly monodisperse Pd-Ni@PVP NPs have been performed by using UV-VIS spectroscopy. The UV-VIS results (Fig. 1) demonstrates the conversion of Pd and Ni salts to Pd-Ni nanomaterials due to the absence of d–d transitions belonging to Pd^{2+} and Ni^{2+} ions.

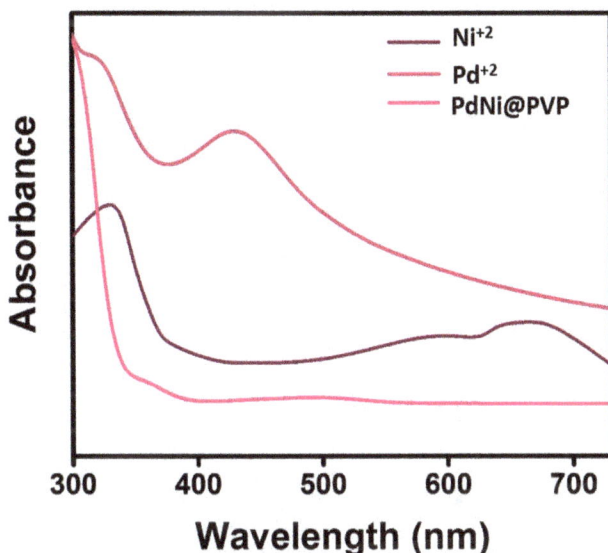

Fig.1 UV-Vis absorption spectra of the aqueous solutions of Pd+2, Ni+2, and Pd-Ni@PVP NPs.

After preparation of Pd-Ni@PVP NPs, their performances were examined for DMAB dehydrocopuling and high efficiency and durability of the uniformly dispersed Pd-Ni@PVP nanoclusters as catalysts has been shown for the dehydrocoupling of DMAB. The details about the dehydrogenation system and procedure was given in the supporting information. The DMAB dehydrogenation was indicated in Fig. 2a in the presence of prepared catalyst with various amounts at 25.0 ±0.1 °C. As shown in this figure, hydrogen evolution begins suddenly and continues until the finishing of the dehydrocoupling of DMAB. The complete conversion were proved by NMR data of DMAB (δ= ~12.7 ppm) to metaborate (δ= ~5 ppm) which shows the conversion of DMAB (at 1.0 equiv. H_2 generation) even at room temperature. As indicated in Fig. 2b, the dehydrocoupling of DMAB with prepared nano catalyst were obtained at various temperatures. This figure was used to determine the rate contants for the dehydrocoupling of DMAB at four different temperatures. The rate constants were also applied to find E_a as 37.11 ± 2 kJ/mol from the Arrhenius graph (Fig. 2c). On the other hand, activation enthalpy ($\Delta H^{\#}$= 34.43 kJ mol^{-1}) and activation entropy ($\Delta S^{\#}$= -92.36 J mol^{-1} K^{-1})) were also calculated by using rate constants from the Eyring plot (Fig. 2d) for dehydrocoupling reaction.

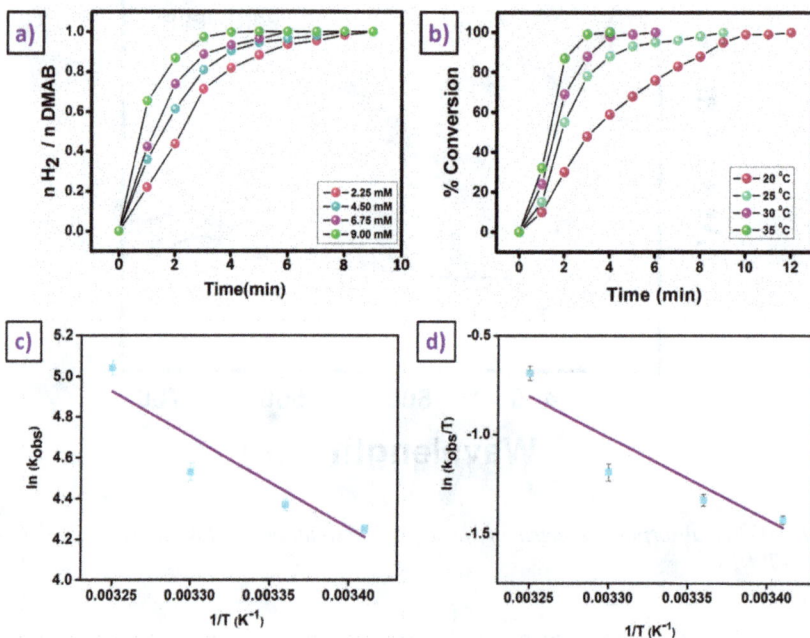

Fig. 2 (a) Plot of nH2/nDMAB versus time(s) in different catalyst concentrations at 25.0 ± 0.1 oC (b) Plots of % conversion versus time graph for catalyst (7.5% mol) in THF at various temperatures (c) Arrhenius and (d) Eyring plots for PdNi@PVP NPs catalysed dehydrocoupling of DMAB at various temperatures.

As a result of catalytic performance, it is valuable to state that the prepared monodisperse catalyst Pd-Ni@PVP nanocomposites entirely evolve hydrogen gas (1.0 mol H_2/mol DMAB) in a very short time with high TOF number of 561.0 h^{-1} at 25.0 ± 0.1 °C, in the catalytic dehydrogenation of DMAB. It is obvious that the prepared catalyst shows very high catalytic performance (561.0 h^{-1}) for DMAB dehydrocoupling as shown in Table 1. The higher catalytic performance of newly prepared, highly monodisperse nanocatalyst Pd-Ni@PVP is most probably due to adequate durability of PVP, and cooperative and synergistic impact of Pd and Ni atoms which is deposited on PVP in a prepared catalyst system that results in the decreased particle magnitude of the catalyst. As a result, the catalytic performance and monodispersity of newly produced Pd-Ni@PVP NPs were efficiently improved by deposition of Pd and Ni nanoparticles on PVP for DMAB dehydrocoupling reaction.

Table 1   The data of turnover frequencies belonging to different investigations during DMAB dehydrogenation.

Entry	(Pre) Catalysts	Conv. (%)	TOF	Ref
1	**PdNi@PVP**	**100**	**561.0**	**This study**
2	RhCl$_3$	90	7.9	11
3	Pd/C	95	2.8	11
4	trans-RuMe$_2$(PMe$_3$)$_4$	100	12.4	11
5	Cp$_2$Ti	100	12.3	30
6	[Ir(1,5-cod)m-Cl]$_2$	95	0.7	11
7	[Rh(1,5-cod)(dmpe)]PF$_6$	95	1.7	11
8	[Rh(1,5-cod)m-Cl]$_2$	100	12.5	11
9	[Rh(1,5-cod)$_2$]Otf	95	12.0	11
10	[RuH(PMe$_3$)(NC2H4PPr$_2$)$_2$]	100	1.5	26
11	IrCl$_3$	25	0.3	11
12	[Cr(CO)$_5$(thf)]	97	13.4	3
13	Rh(0)/[Noct$_4$]Cl	90	8.2	11
14	[RhCl(PHCy$_2$)$_3$]	100	2.6	29
15	RhCl(PPh$_3$)$_3$	100	4.3	11
16	(Idipp)CuCl	100	0.3	8
18	Ni(skeletal)	100	3.2	15
19	[Cp*Rh(m-Cl)Cl]$_2$	100	0.9	25
20	[Cr(CO)$_5$(n^1-BH3NMe$_3$)]	97	19.9	3
21	Ru(cod)(cot)	40	1.6	2

Besides, the reusability of PVP-promoted Pd-Ni nanomaterials was explored as shown in Fig. 3 for DMAB dehydrogenation reactions. The detail about the procedure was given in the supporting information. The PVP-supported Pd-Ni nanocatalyst maintains the catalytic function (80 % of its beginning performance) for the dehydrocoupling of DMAB reaction, after the 4th experiment. The catalytic activity of Pd-Ni@PVP NPs were decreased after the fourth cycle of the experiment as compared to the initial performance which is most probably due to the reducing of the nanoparticles'surface area by increasing the amount of the product, which decreases accessibility of active sites and the aggregation of nanoparticles. Even after the forth cycle the catalyst retains its initial content as found by the ICP (13.44 % metals based) study showing the same effect as it was at the beginning.

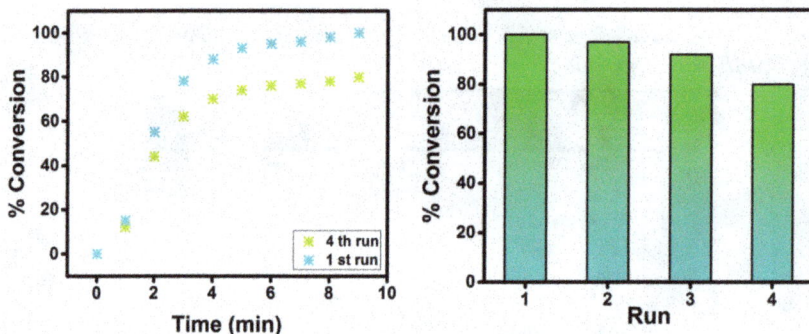

Fig. 3 Plots % conversion versus time graph for PdNi@PVP NPs (7.5 % mol) catalysed dehydrocoupling of DMAB in THF at room temperature for 1ˢᵗ and 4ᵗʰ catalytic runs.

To understand the reason behind the enhanced activity and stability of PdNi@PVP NPs, prepared samples of solid catalyst were characterized by TEM, HRTEM, XRD and XPS.

Firstly, the surface morphology, micro-structural features, and composition of Pd-Ni@PVP NPs were explored by TEM analyses as indicated in Fig. 4. The average particle size of the prepared well dispersed nanoclusters was found to be 3.05 ± 0.38 nm as shown in the particle size histogram. The morphology of the prepared nanoparticles was also illustrated with high resolution electron micrograph (HRTEM) in Fig. 4. As shown in Fig. 4a, the nanoparticles of Pd and Ni were uniformly dispersed on the PVP with a great spherical shape which helped to prevent agglomeration in the synthesized catalyst. Furthermore, the atomic lattice fringes were also observed HRTEM image for monodisperse Pd-Ni@PVP NPs as seen in Fig. 4. As a consequence of these fringes, PdNi (111) were found with plane spacing of 0.21 nm on the prepared catalyst, which is a bit smaller compared to the nominal Pd (111) spacing of 0.22 nm. In addition, EELS spectrum of Pd-Ni@PVP NPs validate the alloy structure and 1:1 ratio of Pd:Ni. The 1:1 ratio of Pd:Ni was also confirmed by the ICP analysis ($Pd_{54}Ni_{46}$).

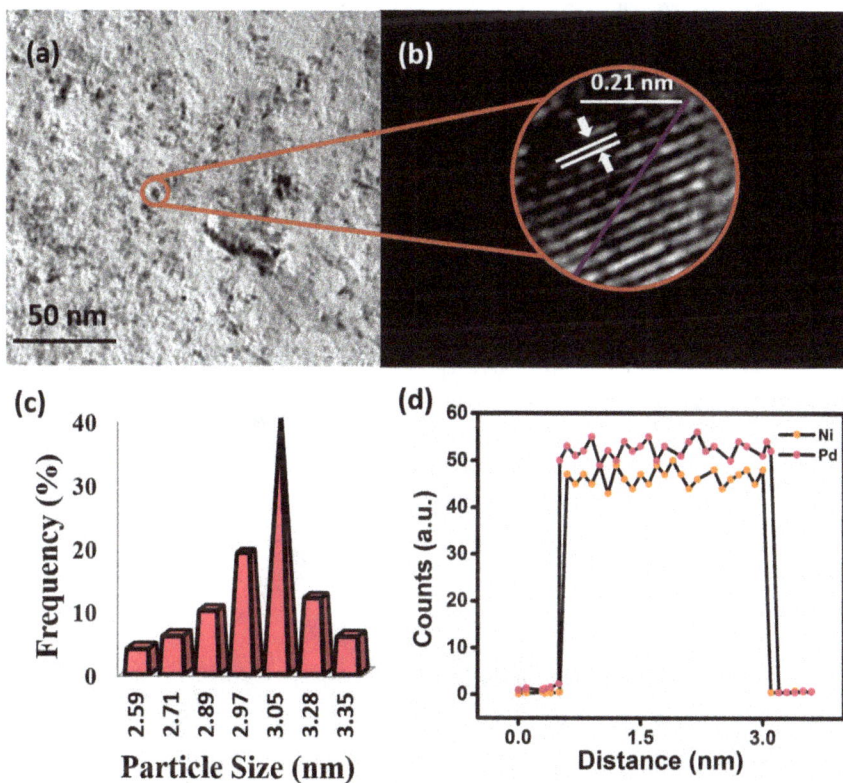

Fig. 4 (a) Transition electron micrograph (b) particle size histogram (c) and HR-TEM image (d) the EELS line profile scanned on the PdNi NPs shown in HRTEM.

The Pd-Ni nanoparticles furnished on PVP were further analyzed by using XRD data to define crystal structure and the average crystallite size of nanoparticles. The Fig. 5 illustrates the XRD pattern of the Pd@PVP and Pd-Ni@PVP as catalysts in order to show the structure of the catalysts. The slight diffraction peak shift was observed from lower 2θ to higher 2θ values for the catalyst containing palladium and nickel as compared the pure palladium, which shows the formation of Pd-Ni@PVP NPs alloy as indicated in Fig. 5. Further, the peak at around 13.8° Is attributed to the PVP and the any characteristic diffraction peak was observed from the XRD pattern for prepared nanocluster because of the relatively strong signal for the Pd species. The four reflections at $2\theta = 40.1°$, $46.5°$, $68.5°$ and $82.6°$ correspond to Pd (111), (200), (220) and (311) are determined as face centered cubic structure of palladium-nickel alloy as clearly seen in Fig. 5. On the other hand, the average crystallite particle size of palladium-nickel deposited PVP NPs have been calculated as about 3.02 ± 0.42 nm using the Scherrer's equation which is in good agreement with TEM analyses results [12, 19, 28].

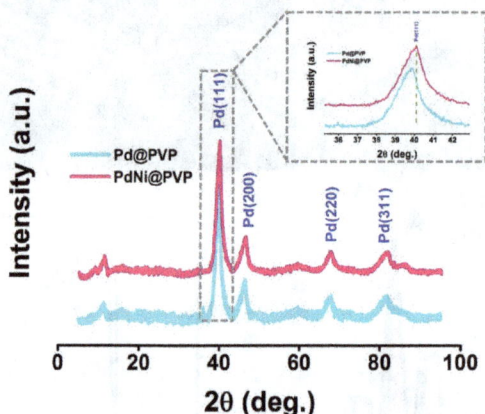

Fig. 5 XRD of Pd@PVP and PdNi@PVP NPs.

To calculate the lattice parameter (αPdNi) values, the PdNi (220) diffraction peak of Pd-Ni@PVP NPs was used. The lattice parameter was calculated as 3.88 Å with the help of the following equation which is a bit smaller compared to pure Pd (3.89 Å) [12].

$$Sin\ \theta = \frac{\lambda\sqrt{h^2+k^2+l^2}}{2a}(\text{for a cubic structure})$$

X-ray photoelectron spectroscopy (XPS) was used to evaluate elemental composition and chemical oxidation states of Pd and Ni present in highly uniform Pd-Ni@PVP nanocomposites. For this aim, the Pd 3d and Ni 2p region of the spectrum was determined and fitted by the Gaussian-Lorentzian method [12, 19, 28]. In Fig. 6, the experimental binding energies (Pd-3d$_{5/2}$: 335.4 eV, Ni-2p$_{3/2}$: 853.8 eV) showed that the surface of Pd and Ni are mostly metallic and not oxides due to PVP wrapping of Pd (0) and Ni(0) metals in the catalyst producing part. The shift of the 2p$_{3/2}$ peak of the binding energy of Ni and lower energy indicates alloying of Ni with Pd. As shown in Fig. 6, the

Pd (II) and Ni (II) peaks may be caused by the surface oxidation and/or chemisorption of environmental oxygen during the preparation process of dehydrogenation.

After the fully characterization of PdNi@PVP NPs by TEM, HRTEM, XRD and XPS, the enhanced activity and stability of the prepared samples of solid catalyst is most probably due to the ultrasmall size, surface area, monodispersity, alloy structure and high % metal ratio of PdNi@PVP NPs.

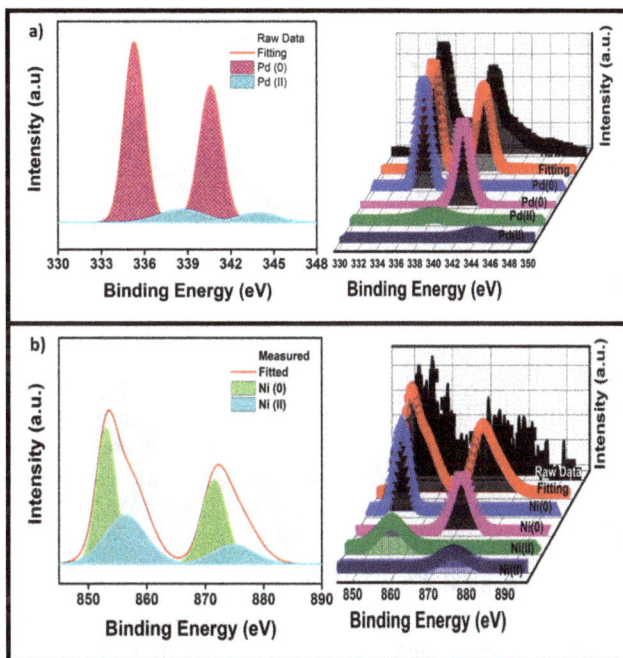

Fig. 6 Pd 3d (a) and Ni 2p (b)XPS spectra of PdNi@PVP NPs.

4. Conclusions

As a conclusion, highly efficient, reusable and stable Pd-Ni@PVP nanoclusters were fabricated with uniform distribution of Pd-Ni NPs on PVP by using the microwave assistance method and they were successfully applied for the catalytic dehydrogenation of DMAB. The Pd-Ni@PVP nanocomposites have been shown to have a high catalytic performance by having a turnover frequency (TOF) of 561.0 h^{-1} in the dehydrogenation of DMAB. On the other hand, the Ea value was also obtained as a 37.11 ± 2 kJ/mol for the catalytic dehydrocoupling of DMAB in the presence of prepared Pd-Ni@PVP nanoclusters. Besides their high catalytic performance, the Pd-Ni@PVP NPs were prepared in a facile, controllable and simple way which can be employed for a broad range of applications in hydrogen storage and fuel cells. Furthermore, this simple and efficient method can be extended to other polymer supported bimetallic or polymetallic nanograins for other applications.

Acknowledgements

The authors would like to thank to DPU-BAP (2014-05, 2015-50) for the financial support.

References

[1]   Q. Zhang, G.M. Smith, Y. Wu, Catalytic hydrolysis of sodium borohydride in an integrated reactor for hydrogen generation, Int. J. Hydrogen Energy 32 (2007) 4731–4735. https://doi.org/10.1016/j.ijhydene.2007.08.017

[2]   M. Zahmakıran, M. Tristany, K. Philippot, K. Fajerweg, S. Özkar, B. Chaudret, Aminopropyltriethoxysilane stabilized ruthenium(0) nanoclusters as an isolable and reusable heterogeneous catalyst for the dehydrogenation of dimethylamine–borane, Chem. Commun. 46 (2010) 2938–2940. https://doi.org/10.1039/c000419g

[3]   Y. Kawano, M. Uruichi, M. Shiomi, S. Taki, T. Kawaguchi, T. Kakizawa, H. Ogino, Dehydrocoupling reactions of borane–secondary and –primary amine adducts catalyzed by group-6 carbonyl complexes, formation of aminoboranes and borazines, J. Am. Chem. Soc. 131 (2009) 14946. https://doi.org/10.1021/ja904918u

[4]   M. Zahmakıran, S. Ozkar, Dimethylammonium hexanoate stabilized rhodium(0) nanoclusters identified as true heterogeneous catalysts with the highest observed activity in the dehydrogenation of dimethylamine–borane, Inorg. Chem. 48 (2009) 8955–8964. https://doi.org/10.1021/ic9014306

[5]   (a) V. Malgras, H. Ataee-Esfahani, H. Wang, B. Jiang, C. Li, K.C.W. Wu, J.H. Kim, Y. Yamauchi, Nanoarchitectures for mesoporous metals, advanced

materials, 28(6) (2016) 993–1010. (b) A. Mittal, J. Mittal, A. Malviya, D. Kaur, V.K. Gupta, Decoloration treatment of a hazardous triarylmethane dye, Light Green SF (Yellowish) by waste material adsorbents, J. Colloid Interface Sci. 342(2) (2010) 518–527. (c) T. Saleh, V.K. Gupta, Photo-catalyzed degradation of hazardous dye methyl orange by use of a composite catalyst consisting of multi-walled carbon nanotubes and titanium dioxide, J. Colloid Interface Sci. 371 (2012) 101–106. (d) A. Mittal, J. Mittal, A. Malviya, V.K. Gupta, Removal and recovery of chrysoidine Y from aqueous solutions by waste materials, J. Colloid Interface Sci. 344(2) (2010) 497–507. https://doi.org/10.1016/j.jcis.2011.12.038

[6]   (a) S.C. Amendola, J.M. Janjua, N.C. Spencer, M.T. Kelly, P.J. Petillo, S.L. Sharp–Goldman, M. Binder, A safe, portable, hydrogen gas generator using aqueous borohydride solution and Ru catalyst, Int. J. Hydrogen Energy 25 (2000) 969–975. (b) A. Mittal, J. Mittal, A. Malviya, V.K. Gupta, Adsorptive removal of hazardous anionic dye "Congo red" from wastewater using waste materials and recovery by desorption, J. Colloid Interface Sci. 340 (2009) 16–26. (c) A. Mittal, D. Kaur, J. Mittal, A. Malviya, V.K. Gupta, Adsorption studies on the removal of coloring agent phenol red from wastewater using waste materials as adsorbents, J. Colloid Interface Sci. 337 (2009) 345–354. https://doi.org/10.1016/j.jcis.2009.05.016

[7]   B. Sen, S. Kuzu, E. Demir, E. Yıldırır, F. Sen, Highly efficient catalytic dehydrogenation of dimethly ammonia borane via monodisperse palladium-nickel alloy nanoparticles assembled on PEDOT, Int. J. Hydrogen Energy (2017). https://doi.org/10.1016/j.ijhydene.2017.05.115.

[8]   R.J. Keaton, J.M. Blacquiere, R.T. Baker, Base metal catalyzed dehydrogenation of ammonia–borane for chemical hydrogen storage, J. Am. Chem. Soc. 129 (2007) 11936. https://doi.org/10.1021/ja066860i

[9]   N. Mohajeri, A. T-Raissi, O. Adebiyi, Hydrolytic cleavage of ammonia-borane complex for hydrogen production, J. Power Sources 167 (2007) 482–485. https://doi.org/10.1016/j.jpowsour.2007.02.059

[10]   P.V. Ramachandran, P.D. Gagare, Preparation of ammonia borane in high yield and purity, methanolysis, and regeneration, Inorg. Chem. 46 (2007) 7810–7817. https://doi.org/10.1021/ic700772a

[11]   C.A. Jaska, I. Manners, Heterogeneous or homogeneous catalysis? mechanistic studies of the rhodium-catalyzed dehydrocoupling of amine-borane and

phosphine-borane adducts, J. Am. Chem. Soc. 126 (2004) 9776–9785.
https://doi.org/10.1021/ja0478431

[12] B. Çelik, Y. Yıldız, H. Sert, E. Erken, Y. Koşkun, F. Sen, Monodispersed palladium–cobalt alloy nanoparticles assembled on poly(N-vinyl-pyrrolidone) (PVP) as a highly effective catalyst for dimethylamine borane (DMAB) dehydrocoupling, RSC Adv. 6 (2016) 24097–24102.
https://doi.org/10.1039/C6RA00536E

[13] Q. Xu, M. Chandra, A portable hydrogen generation system, catalytic hydrolysis of ammonia–borane, J. Alloy Compd. 446–447 (2007) 729–732.
https://doi.org/10.1016/j.jallcom.2007.01.040

[14] Q. Xu, M. Chandra, Catalytic activities of non-noble metals for hydrogen generation from aqueous ammonia–borane at room temperature, J. Power Sources, 163 (2006) 364–370. https://doi.org/10.1016/j.jpowsour.2006.09.043

[15] A.P.M. Robertson, R. Suter, L. Chabanne, G.R. Whittel, I. Manners, Heterogeneous dehydrocoupling of amine–borane adducts by skeletal nickel catalysts, Inorg. Chem. 50 (2011) 12680. https://doi.org/10.1021/ic201809g

[16] T. Umegaki, J.M. Yan, X.B. Zhang, H. Shioyama, N. Kuriyama, Q. Xu, Preparation and catalysis of poly(N-vinyl-2-pyrrolidone) (PVP) stabilized nickel catalyst for hydrolytic dehydrogenation of ammonia borane, Int. J. Hydrogen Energy 34 (2009) 3816–3822.
https://doi.org/10.1016/j.ijhydene.2009.03.003

[17] R. Fernandes, N. Patel, A. Miotello, Hydrogen generation by hydrolysis of alkaline NaBH4 solution with Cr-promoted Co–B amorphous catalyst, Appl. Catal. B. Environ. 92 (2009) 68–74.
https://doi.org/10.1016/j.apcatb.2009.07.019

[18] J.M. Yan, X.B. Zhang, S. Han, H. Shioyama, Q. Xu, Iron-nanoparticle-catalyzed hydrolytic dehydrogenation of ammonia borane for chemical hydrogen storage, Angew.Chem. Int. Ed. 47 (2008) 2287–2289.
https://doi.org/10.1002/anie.200704943

[19] B. Sen, S. Kuzu, E. Demir, T.O. Okyay, F. Sen, Hydrogen liberation from the dehydrocoupling of dimethylamine-bor.ane at room temperature by using novel and highly monodispersed RuPtNi nanocatalysts decorated with graphene oxide, Int. J. Hydrogen Energy (2017).
https://doi.org/10.1016/j.ijhydene.2017.04.213

[20] J.M. Yan, X.B. Zhang, H. Shioyama, Q. Xu, Room temperature hydrolytic dehydrogenation of ammonia borane catalyzed by Co nanoparticles, J. Power Sources 195 (2010) 1091–1094. https://doi.org/10.1016/j.jpowsour.2009.08.067

[21] J.M. Yan, X.B. Zhang, S. Han, H. Shioyama, Q. Xu, Synthesis of longtime water/air-stable Ni nanoparticles and their high catalytic activity for hydrolysis of ammonia–borane for hydrogen generation, Inorg. Chem. 48 (2009) 7389–7393. https://doi.org/10.1021/ic900921m

[22] D.G. Tong, X.L. Zeng, W. Chu, D. Wang, P. Wu, Magnetically recyclable hollow Co–B nanospindles as catalysts for hydrogen generation from ammonia borane, J. Mater. Sci. 45 (2010) 2862–2867. https://doi.org/10.1007/s10853-010-4275-0

[23] N. Patel, R. Fernandes, G. Guella, A. Miotello, Nanoparticle-assembled Co-B thin film for the hydrolysis of ammonia borane: A highly active catalyst for hydrogen production, Appl. Catal. B Environ. 95 (2010) 137–143. https://doi.org/10.1016/j.apcatb.2009.12.020

[24] Y. Yamada, K. Yano, Q. Xu, S. Fukuzumi, Cu/Co3O4 nanoparticles as catalysts for hydrogen evolution from ammonia borane by hydrolysis, J. Phys. Chem. C 114 (2010) 16456–16462. https://doi.org/10.1021/jp104291s

[25] C.A. Jaska, K. Temple, A.J. Lough, I. Manners, Transition metal-catalyzed formation of boron–nitrogen bonds: catalytic dehydrocoupling of amine-borane adducts to form aminoboranes and borazines, J. Am. Chem. Soc. 125 (2003) 9424–9434. https://doi.org/10.1021/ja0301601

[26] A. Friederich, M. Drees, S. Schneider, Ruthenium-catalyzed dimethylamineborane dehydrogenation, stepwise metal-centered dehydrocyclization, Chem. Eur. J. 15 (2009) 10339–10342. https://doi.org/10.1002/chem.200901372

[27] M. Chandra, Q. Xu, A high-performance hydrogen generation system: Transition metal-catalyzed dissociation and hydrolysis of ammonia–borane, J. Power Sources 156 (2006) 190–194. https://doi.org/10.1016/j.jpowsour.2005.05.043

[28] Y. Chen, J.L. Fulton, J.C. Linehan, T. Autrey, In Situ XAFS and NMR Study of Rhodium-Catalyzed dehydrogenation of dimethylamine borane, J. Am. Chem. Soc. 127 (2005) 3254–3255. https://doi.org/10.1021/ja0437050

[29] M.E. Sloan, A. Staubitz, T.J. Clark, C.A. Russell, G.C. Lloyd-Jones, I.
      Manners, Homogeneous catalytic dehydrocoupling/dehydrogenation of
      amine−borane adducts by early transition metal, group 4 metallocene
      complexes, J. Am. Chem. Soc. 132 (2010) 3831–3841.
      https://doi.org/10.1021/ja909535a

[30] T.J. Clark, C.A. Russell, I. Manners, Homogeneous, Titanocene-catalyzed
      dehydrocoupling of amine−borane adducts, J. Am. Chem. Soc. 128 (2006)
      9582–9583. https://doi.org/10.1021/ja062217k

[31] G. Alcaraz, L. Vendier, E. Clot, S. Sabo-Etienne, Ruthenium Bis(σ-B-H)
      aminoborane complexes from dehydrogenation of amine−boranes, Trapping of
      H2B-NH2, Angew. Chem. Int. Ed. 49 (2010) 918–920.
      https://doi.org/10.1002/anie.200905970

[32] Y. Yildiz, E. Erken, H. Pamuk, H. Sert, F. Sen, Monodisperse Pt nanoparticles
      assembled on reduced graphene oxide, highly efficient and reusable catalyst
      for methanol oxidation and dehydrocoupling of dimethylamine-borane
      (DMAB), J. Nanosci. Nanotech. 6 (2016) 5951–5958.
      https://doi.org/10.1166/jnn.2016.11710

[33] E. Erken, Y. Yildiz, B. Kilbas, F. Sen. Synthesis and Characterization of
      Nearly Monodisperse Pt Nanoparticles for C1 to C3 Alcohol Oxidation and
      Dehydrogenation of Dimethylamine-borane (DMAB), J. Nanosci.
      Nanotechnol. 16 (2016) 5944–5950. https://doi.org/10.1166/jnn.2016.11683

[34] B. Çelik, G. Başkaya, Ö. Karatepe, E. Erken, F. Sen. Monodisperse
      Pt(0)/DPA@GO Nanoparticles as Highly Active Catalysts for Alcohol
      Oxidation and Dehydrogenation of DMAB, Int. J. Hydrogen Energy 41 (2016)
      5661–5669. https://doi.org/10.1016/j.ijhydene.2016.02.061

[35] Y. Jiang, H. Berke, Dehydrocoupling of dimethylamine-borane catalysed by
      rhenium complexes and its application in olefin transfer-hydrogenations,
      Chem. Commun. 34 (2007) 3571–3573. https://doi.org/10.1039/b708913a

[36] JL Fulton, JC Linehan, T. Autrey, M. Balasubramanian, Y. Chen, NK.
      Szymczak, When is a nanoparticle a cluster? an operando EXAFS study of
      amine borane dehydrocoupling by Rh4-6 Clusters, J. Am. Chem. Soc. 129
      (2007) 11936–11949. https://doi.org/10.1021/ja0733311

[37] M.E. Sloan, T.C. Clars, I. Manners, Homogeneous catalytic
      dehydrogenation/dehydrocoupling of amine-borane adducts by the Rh(I)

Wilkinson's complex analogue RhCl(PHCy2)3 (Cy = cyclohexyl), Inorg. Chem, 48 (2009) 2429–2435. https://doi.org/10.1021/ic801752k

[38]   B. Celik, S. Kuzu, E. Erken, H. Sert, Y. Koskun, F. Sen, Nearly monodisperse carbon nanotube furnished nanocatalysts as highly efficient and reusable catalyst for dehydrocoupling of DMAB and C1 to C3 alcohol oxidation, Int. J. Hydrogen Energy 41 (2016) 3093–3101. https://doi.org/10.1016/j.ijhydene.2015.12.138

[39]   M. Munoz-Olasagasti, A. Telleria, J. Perez-Miqueo, M.A. Garralda, Z.A. Freixa, A readily accessible ruthenium catalyst for the solvolytic dehydrogenation of amine–borane adducts, Dalton Trans, 43 (2014) 11404–11409. https://doi.org/10.1039/c4dt01216j

[40]   F. Sen, Karatas Y, Gulcan M, Zahmakiran M, Amylamine stabilized platinum(0) nanoparticles: active and reusable nanocatalyst in the room temperature dehydrogenation of dimethylamine-borane, RSC Adv. 4 (2014) 1526–1531. https://doi.org/10.1039/C3RA43701A

[41]   E. Erken, H. Pamuk, Ö. Karatepe, G. Başkaya, H. Sert, M. O. Kalfa, F. Sen, New Pt(0) nanoparticles as highly active and reusable catalysts in the C1–C3 alcohol oxidation and the room temperature dehydrocoupling of dimethylamine-borane (DMAB), J. Cluster Sci. 27(9) (2016) 23. https://doi.org/10.1007/s10876-015-0892-8

[42]   B. Çelik, E. Erken, S. Eriş, Y. Yıldız, B. Şahin, H. Pamuk, F. Sen, Highly monodisperse Pt(0)@AC NPs as highly efficient and reusable catalysts: the effect of the surfactant on their catalytic activities in room temperature dehydrocoupling of DMAB, Catal. Sci. Technol. 6 (2016) 1685–1692. https://doi.org/10.1039/C5CY01371B

[43]   B. Celik, S. Kuzu, E. Erken, H. Sert, Y. Koskun, F. Sen. Nearly Monodisperse Carbon Nanotube Furnished Nanocatalysts as Highly Efficient and Reusable Catalyst for Dehydrocoupling of DMAB and C1 to C3 Alcohol Oxidation, Int. J. Hydrogen Energy 41 (2016) 3093–3101. https://doi.org/10.1016/j.ijhydene.2015.12.138

[44]   D. Pun, E. Lobkovsky, P.J. Chirik, Amineborane dehydrogenation promoted by isolable zirconium sandwich, titanium sandwich and N2 complexes, Chem. Commun. 44 (2007) 3297. https://doi.org/10.1039/b704941b

[45] C.F. Yao, L. Zhuang, Y.L. Cao, X.P. Hi, H.X. Yang, Hydrogen release from hydrolysis of borazane on Pt- and Ni-based alloy catalysts, Int. J. Hydrogen Energy 33 (2008) 2462–2467. https://doi.org/10.1016/j.ijhydene.2008.02.028

[46] X. Yang, F. Cheng, J. Liang, Z. Tao, J. Chen, PtxNi1−x nanoparticles as catalysts for hydrogen generation from hydrolysis of ammonia borane, Int. J. Hydrogen Energy 34 (2009) 8785–8791. https://doi.org/10.1016/j.ijhydene.2009.08.075

[47] H.B. Dai, L.L. Gao, Y. Liang, X.D. Kang, P. Wang, Promoted hydrogen generation from ammonia borane aqueous solution using cobalt–molybdenum–boron/nickel foam catalyst, J. Power Sources 195 (2010) 307–312. https://doi.org/10.1016/j.jpowsour.2009.06.094

[48] B. Sen, S. Kuzu, E. Demir, S. Akocak, F. Sen, Monodisperse palladium-nickel alloy nanoparticles assembled on graphene oxide with the high catalytic activity and reusability in the dehydrogenation of dimethylamine-borane, Int. J. Hydrogen Energy (2017) https://doi.org/10.1016/j.ijhydene.2017.05.113.

[49] B. Sen, S. Kuzu, E. Demir, S. Akocak, F. Sen, Polymer-Graphene hybride decorated Pt Nanoparticles as highly efficient and reusable catalyst for the Dehydrogenation of Dimethylamine-borane at room temperature, Int. J. Hydrogen Energy (2017). https://doi.org/10.1016/j.ijhydene.2017.05.112.

[50] B. Sen, S. Kuzu, E. Demir, S. Akocak, F. Sen, Highly monodisperse RuCo nanoparticles decorated on functionalized multiwalled carbon nanotube with the highest observed catalytic activity in the dehydrogenation of dimethylamine borane, Int. J. Hydrogen Energy (2017). https://doi.org/10.1016/j.ijhydene.2017.06.032.

Keyword Index

About the Editor

Dr. Abu Nasar is presently working as an Associate Professor at the Department of Applied Chemistry, Faculty of Engineering & Technology, Aligarh Muslim University.

He has received his PhD and done postdoctoral work at the prestigious organization, Indian Institute of Technology, Banaras Hindu University. He has published over three dozen research papers in reputed journals.

His areas of interest include physical chemistry, environmental chemistry and materials science.

A research paper presented by him at the International Conference on "Advanced Semiconductor Devices and Microsystems" held at Smolenice, Slovakia during Oct. 20-24, 1996 was selected as one of the **best papers** of the Conference. In the recognition of his work, Dr. Nasar has received the **Young Scientists'** and **Young Metallurgists' Awards** by the Indian Science Congress Association and the Ministry of Steel, Government of India, respectively.

www.ingramcontent.com/pod-product-compliance
Lightning Source LLC
Chambersburg PA
CBHW061020220326
41597CB00016BB/1761